FEATHERED WITH
Puzzled
Parrot
MIND-STIMULATING CHALLENGES

LARGE PRINT
NUMBER
SEARCH
PUZZLES BOOK FOR ADULTS
VOLUME 1

THIS BOOK BELONGS TO

_____

PUZZLED PARROT

CONTENTS

PUZZLED P PARROT

How To Play Number Search

Number Search challenges players to identify and mark all hidden number sequences within the given set of numbers in a rectangular or square grid.

These sequences can be horizontal, vertical, diagonal, forward or backward, and the numbers themselves can be any direction. Typically, a list of hidden number sequences is given below the puzzle for reference.

8	4	3	0	5	1	4	9	1	0
5	4	3	9	8	3	7	0	6	5
6	2	3	8	5	8	8	4	7	1
2	0	3	4	5	5	0	8	3	0
3	8	2	8	1	9	8	7	6	2
0	8	2	3	9	1	2	7	8	6
1	8	6	5	8	5	7	4	5	3
2	7	3	3	5	0	7	1	2	6
2	5	2	0	5	9	0	2	2	1
9	1	0	2	0	7	6	9	2	9

8	4	3	0	5	1	4	9	1	0
5	4	3	9	8	3	7	0	6	5
6	2	3	8	5	8	8	4	7	1
2	0	3	4	5	5	0	8	3	0
3	8	2	8	1	9	8	7	6	2
0	8	2	3	9	1	2	7	8	6
1	8	6	5	8	5	7	4	5	3
2	7	3	3	5	0	7	1	2	6
2	5	2	0	5	9	0	2	2	1
9	1	0	2	0	7	6	9	2	9

24295	523895
6073893	748858
5205	33507
1711434	5785
221032	62233

24295	523895
6073893	748858
5205	33507
1711434	5785
221032	62233

Dear Valued Puzzler,

Thank you for purchasing our number search puzzle book!

If you enjoyed it, we would greatly appreciate it if you could leave a review on Amazon. Your feedback is invaluable in helping others and allowing us to continue creating high-quality products.

Thank you for your support!

PUZZLED · PARROT

Puzzle 1

```
1 1 1 4 7 7 0 6 3 2 4 7 1 6 3 7
9 2 2 0 9 7 8 8 6 4 7 8 8 2 1 5
2 3 7 3 4 4 5 2 2 0 4 8 6 1 8 2
6 9 5 4 7 6 7 7 0 2 2 5 2 5 6 7
9 9 6 1 6 5 2 6 5 5 2 6 0 8 2 3
6 4 1 9 6 2 9 1 4 6 0 0 0 4 0 4
8 2 5 8 6 7 5 0 0 8 6 7 6 3 9 7
8 1 9 3 3 3 5 9 9 7 3 2 6 4 0 8
8 5 5 7 8 9 7 4 0 6 9 6 6 6 5 1
4 0 9 6 7 8 7 9 7 8 6 1 2 5 6 9
5 5 6 4 3 7 7 8 2 1 5 7 7 8 4 4
2 3 7 8 2 1 6 4 7 3 5 8 4 4 2 8
6 5 4 8 3 2 7 1 6 9 8 1 4 8 2 2
9 6 8 3 0 2 2 7 3 1 1 1 0 5 5 9
1 5 5 2 2 0 7 9 0 5 6 1 8 4 5 3
9 1 2 7 2 3 0 1 6 9 5 5 3 4 9 5
7 3 5 0 2 9 2 3 8 4 1 7 3 2 7 8
1 5 6 7 0 4 9 3 6 1 5 2 4 3 5 4
```

- [ ] 792381
- [ ] 167547
- [ ] 967485
- [ ] 002681
- [ ] 742206
- [ ] 926968
- [ ] 384173
- [ ] 207767
- [ ] 051249
- [ ] 096674
- [ ] 846738
- [ ] 756159
- [ ] 032721
- [ ] 962548
- [ ] 688790
- [ ] 197148

Puzzle 2

```
2 1 7 0 7 1 1 0 5 1 8 0 3 2 7 2
4 9 8 3 8 4 5 5 3 1 8 1 2 7 0 3
3 7 6 2 5 0 9 7 5 4 7 0 7 7 0 2
5 2 1 0 8 1 1 2 4 3 3 3 9 8 0 3
4 6 4 6 1 3 0 8 5 6 8 0 9 2 7 9
4 1 0 6 8 9 4 3 5 1 0 6 3 8 6 0
8 4 4 8 3 5 3 7 2 0 4 1 7 7 4 7
1 0 4 3 2 2 4 2 4 4 9 3 3 5 6 6
3 7 3 6 0 1 2 3 3 4 9 7 1 2 3 6
7 5 4 6 8 4 6 4 8 9 6 4 9 8 3 6
6 7 2 5 7 3 8 6 6 7 8 6 2 8 2 2
6 6 5 1 5 4 6 1 2 3 5 2 9 3 3 5
9 8 2 7 8 3 0 8 0 4 1 4 9 5 0 1
6 8 7 8 1 7 8 4 0 3 9 8 6 7 8 1
3 2 4 2 5 9 2 6 3 8 4 0 1 7 8 6
0 8 3 6 6 3 3 4 1 7 3 2 0 0 9 0
8 6 9 2 8 5 5 6 6 1 6 9 6 9 7 6
4 9 5 1 6 5 2 1 8 4 0 6 7 4 1 6
```

☐ 672573	☐ 850979	☐ 266128	☐ 355483
☐ 929137	☐ 178343	☐ 000764	☐ 041687
☐ 324259	☐ 683554	☐ 143048	☐ 084264
☐ 487187	☐ 474382	☐ 423015	☐ 633230

Puzzle 3

```
7 5 7 2 6 5 5 3 2 0 9 0 4 0 8 9
8 7 9 8 1 3 5 5 1 5 1 8 3 0 1 9
3 6 9 5 5 3 0 5 0 1 1 9 6 2 3 6
5 8 4 7 3 6 1 7 9 5 1 4 8 8 9 6
3 6 5 3 8 9 6 4 4 8 5 7 1 7 4 7
7 4 7 9 3 1 7 1 1 6 9 6 4 3 8 7
9 3 5 9 8 0 9 0 2 3 1 9 8 9 1 4
4 5 8 5 4 1 7 2 3 3 9 8 2 3 0 4
0 9 9 7 0 4 0 5 5 5 1 6 6 3 0 6
6 8 7 7 5 9 4 6 7 7 2 9 4 5 3 2
5 4 3 1 7 0 2 7 4 5 3 5 6 5 0 7
9 1 7 9 8 3 1 0 5 9 0 6 6 3 0 1
0 6 3 3 6 7 0 1 0 4 2 0 6 7 5 2
6 7 3 2 6 6 7 9 7 1 6 9 5 0 6 1
4 9 1 8 5 1 2 2 6 2 8 2 6 1 2 4
8 1 2 7 0 4 4 4 8 8 2 9 7 3 8
5 6 0 1 1 5 9 4 7 4 5 5 7 1 4 2
2 6 1 9 2 8 7 8 0 5 3 7 0 6 9 9
```

- ☐ 163748
- ☐ 649577
- ☐ 757500
- ☐ 560497
- ☐ 745571
- ☐ 989923
- ☐ 769294
- ☐ 809023
- ☐ 389644
- ☐ 983105
- ☐ 878910
- ☐ 468675
- ☐ 369553
- ☐ 056816
- ☐ 444882
- ☐ 951106

Puzzle 4

```
3 7 9 8 7 6 2 6 7 8 7 2 5 9 9 7
9 7 2 7 5 8 5 9 7 5 8 3 2 6 9 0
5 3 2 2 8 0 7 4 0 6 7 5 9 6 5 9
0 6 6 3 3 7 0 3 6 0 2 6 2 5 9 5
5 4 7 4 2 8 1 9 8 6 4 4 3 0 2 1
5 9 6 7 4 6 8 8 6 8 1 5 9 6 4 5
3 6 6 2 5 7 3 7 8 1 1 3 6 1 7 0
7 8 9 6 9 8 5 4 8 7 2 4 9 1 0 0
2 4 6 1 2 9 9 6 6 5 7 2 4 3 7 7
5 6 4 6 9 8 1 8 2 5 8 8 3 7 9 4
7 4 7 1 7 6 7 1 8 7 9 0 8 5 6 7
2 8 4 4 2 3 1 9 9 0 1 3 7 3 2 3
6 4 2 3 3 7 9 8 7 3 3 8 1 9 3 5
1 2 1 6 2 2 0 9 9 8 1 3 1 5 7 6
2 2 3 1 3 0 2 6 4 8 2 2 9 5 6 4
2 6 0 3 4 6 6 0 6 5 8 9 1 7 0 1
7 6 3 1 0 9 9 4 3 7 7 5 0 7 1 7
1 8 2 1 4 8 1 4 5 7 0 8 0 9 8 1
```

- [ ] 677589
- [ ] 412706
- [ ] 956436
- [ ] 546532
- [ ] 364475
- [ ] 876267
- [ ] 395055
- [ ] 981315
- [ ] 622484
- [ ] 376683
- [ ] 281986
- [ ] 308243
- [ ] 280740
- [ ] 627527
- [ ] 429599
- [ ] 734990

Puzzle 5

```
0 4 6 7 6 3 3 7 7 3 4 6 2 3 0 6
4 5 9 1 7 7 3 3 4 4 6 2 4 5 7 8
7 7 4 0 1 1 3 6 1 5 0 1 1 7 6 9
4 2 2 2 2 9 8 9 6 6 0 1 4 1 7 5
8 2 9 7 7 1 4 0 3 0 1 0 8 4 9 5
4 2 0 1 1 8 4 5 9 9 5 0 8 8 6 0
0 0 2 5 7 3 5 4 6 0 2 1 8 1 9 1
8 5 5 2 8 5 9 9 9 4 0 0 2 3 6 6
3 9 7 4 2 8 5 3 7 2 6 9 7 4 4 6
7 4 3 2 1 6 9 0 8 1 5 4 3 6 0 0
2 8 9 0 5 4 3 4 5 5 9 7 6 8 5 8
8 7 3 3 1 7 4 8 9 0 0 8 2 0 1 9
5 4 1 5 8 6 3 5 1 8 0 7 6 1 8 7
4 1 0 0 2 6 1 4 5 5 6 3 6 4 4 8
3 8 7 4 5 0 9 5 0 9 5 0 4 1 3 0
5 7 2 7 5 9 3 7 4 0 6 9 5 7 6 3
4 8 2 1 6 9 0 5 9 3 4 0 0 6 6 2
7 5 7 3 3 6 4 7 8 8 1 4 3 9 3 0
```

☐ 771809	☐ 098471	☐ 806464	☐ 735460
☐ 345827	☐ 095095	☐ 251006	☐ 375275
☐ 697670	☐ 205889	☐ 385076	☐ 884142
☐ 673939	☐ 542785	☐ 689179	☐ 593400

Puzzle 6

```
9 2 0 0 9 1 1 1 5 5 1 3 1 6 0 6
7 3 0 1 8 6 7 1 1 9 6 0 6 3 4 7
2 4 8 7 5 9 9 9 2 9 0 4 6 7 1 1
2 1 1 1 0 2 9 6 4 3 7 1 0 2 4 4
1 2 7 8 0 4 6 2 4 6 7 5 6 1 1 9
9 9 4 9 2 5 2 4 1 9 6 7 7 6 8 4
0 2 5 2 1 1 3 5 3 5 6 7 6 8 5 7
1 7 7 7 7 9 9 9 8 9 7 5 1 7 3 4
5 8 0 0 9 5 2 3 1 3 0 0 3 3 9 4
2 6 0 6 0 4 7 3 3 7 2 2 3 4 9 5
3 8 1 5 5 5 5 0 2 6 9 3 1 8 4 5
6 4 6 6 5 8 3 3 8 1 0 4 4 2 0 4
6 5 8 2 7 2 3 4 6 9 0 1 5 6 5 8
6 0 3 0 8 8 2 2 7 4 9 2 5 3 9 4
1 3 7 5 2 2 3 8 9 3 2 9 3 9 6 5
5 7 2 2 3 1 1 6 7 7 0 2 9 6 6 2
0 6 5 1 8 9 9 4 4 7 1 3 4 4 4 8
5 4 0 9 8 4 3 7 4 2 9 5 3 0 2 9
```

- ☐ 232919
- ☐ 231167
- ☐ 427570
- ☐ 957945
- ☐ 091115
- ☐ 799623
- ☐ 061315
- ☐ 077667
- ☐ 395963
- ☐ 049935
- ☐ 673054
- ☐ 822550
- ☐ 722190
- ☐ 029009
- ☐ 603088
- ☐ 660676

Puzzle 7

```
3 0 1 2 1 7 7 0 8 0 3 1 9 3 8 3
1 1 4 4 6 0 0 6 9 6 5 3 5 1 3 8
0 9 1 9 7 5 8 0 0 3 2 3 4 5 9 4
2 2 6 9 6 0 1 9 3 7 4 0 9 0 9 3
7 6 3 3 9 4 8 5 3 7 3 4 5 7 4 4
7 8 0 8 5 4 4 9 7 7 5 1 8 2 2 9
8 8 5 2 7 3 0 6 7 3 9 2 3 8 8 0
8 2 1 8 2 0 1 6 6 0 8 8 6 4 1 1
7 2 7 8 1 7 9 5 1 8 6 4 2 7 8 5
1 4 5 7 8 2 0 8 0 8 8 8 2 8 9 5
4 7 2 2 9 6 7 8 7 1 8 6 3 6 1 3
6 9 8 4 2 8 4 1 6 0 6 5 8 4 7 5
0 1 8 2 4 1 1 1 8 7 4 1 1 7 3 0
9 9 0 0 3 4 8 7 1 4 8 7 5 7 3 6
3 0 4 0 4 1 8 1 0 4 9 0 8 7 7 0
5 0 8 4 2 9 2 5 7 8 3 6 7 0 0 3
1 1 7 2 7 8 2 7 1 1 5 6 0 7 1 5
4 0 2 2 8 9 7 1 7 6 8 4 5 9 7 7
```

- ☐ 437358
- ☐ 802875
- ☐ 468806
- ☐ 984284
- ☐ 832937
- ☐ 181404
- ☐ 150728
- ☐ 157794
- ☐ 940618
- ☐ 163681
- ☐ 872468
- ☐ 839130
- ☐ 196353
- ☐ 887720
- ☐ 102812
- ☐ 783670

Puzzle 8

```
9 6 9 9 6 3 8 1 0 4 8 4 3 2 2 9
1 5 2 8 3 3 5 0 5 7 5 4 8 1 4 2
2 8 3 1 6 3 8 0 2 8 7 0 4 6 9 5
1 1 4 2 8 9 1 8 8 5 4 7 8 5 6 6
9 2 4 9 7 2 7 7 0 9 7 4 0 1 2 7
9 9 4 2 0 8 4 5 1 5 9 5 5 9 1 6
7 1 0 0 4 7 1 4 8 7 7 6 0 4 1 7
2 9 0 2 5 4 4 3 6 8 8 3 5 7 4 4
6 0 0 4 5 0 6 4 9 3 1 4 9 2 8 1
4 6 2 9 8 6 3 5 4 8 9 2 9 6 1 2
0 1 7 6 9 7 4 0 4 2 4 4 0 2 3 6
0 7 8 5 7 1 9 6 5 5 1 7 2 6 7 3
8 2 7 5 8 0 0 8 1 2 6 3 1 8 0 1
2 0 3 5 3 6 5 6 7 7 8 9 2 9 2 0
5 5 7 4 4 2 8 0 4 9 2 4 9 6 0 1
9 1 3 2 3 9 0 4 0 4 6 7 4 3 1 2
1 5 7 4 3 7 2 7 2 0 3 4 9 7 6 0
3 2 3 0 8 8 0 3 3 1 8 3 3 4 9 4
```

☐ 008126	☐ 442804	☐ 364147	☐ 749156
☐ 004627	☐ 440745	☐ 040467	☐ 292189
☐ 635302	☐ 824936	☐ 496211	☐ 025646
☐ 457800	☐ 574797	☐ 784019	☐ 050421

Puzzle 9

```
6 9 2 7 3 0 2 2 8 2 8 6 4 6 8 2
0 1 9 4 0 0 3 4 8 6 1 7 2 3 9 3
4 0 0 1 7 9 1 6 2 6 6 2 4 1 2 7
5 1 4 4 6 8 4 4 7 5 0 3 8 2 6 0
2 6 5 8 7 6 6 5 2 4 6 0 5 0 2 5
7 6 4 7 9 8 2 2 1 6 5 5 8 8 0 7
9 2 4 4 3 2 5 3 2 1 2 8 3 6 4 6
4 0 8 4 2 5 0 8 4 1 2 4 8 3 4 6
1 5 3 0 1 1 9 4 8 4 6 4 0 1 3 0
4 3 3 9 1 4 7 1 7 4 1 5 5 6 9 9
1 8 3 9 9 3 1 2 6 0 5 4 2 6 4 1
0 9 2 3 7 3 6 9 1 8 7 6 1 0 4 5
7 2 9 1 7 5 0 4 8 5 3 2 7 8 0 2
1 9 8 1 0 5 0 1 1 2 1 8 1 3 9 9
0 0 5 3 8 3 9 3 6 0 4 6 8 2 4 2
0 6 6 5 6 3 3 2 3 6 3 6 3 9 1 6
6 3 6 0 1 4 5 6 7 9 7 9 8 3 7 1
9 7 9 9 4 6 8 1 8 7 7 1 3 7 0 0
```

- ☐ 662053
- ☐ 866358
- ☐ 489204
- ☐ 432661
- ☐ 844750
- ☐ 376548
- ☐ 062139
- ☐ 174155
- ☐ 791626
- ☐ 301284
- ☐ 930146
- ☐ 256122
- ☐ 858740
- ☐ 014567
- ☐ 463821
- ☐ 305793

10

Puzzle 10

```
7 4 2 2 7 5 2 1 7 3 0 1 8 5 3 2
0 6 4 6 1 9 7 5 6 5 8 9 3 3 6 1
5 6 3 4 1 9 9 9 1 4 7 4 1 6 5 0
2 2 1 7 8 3 6 1 6 4 8 9 8 2 4 8
1 7 0 4 5 9 0 6 6 4 4 8 1 4 6 0
3 2 8 7 8 4 3 0 7 8 1 1 0 0 6 1
8 2 0 2 3 5 7 2 6 0 9 7 4 7 3 3
5 6 5 2 3 9 7 4 1 0 8 2 0 7 3 4
3 4 0 5 6 6 8 6 5 7 1 5 5 1 8 5
5 4 3 7 0 4 4 4 3 7 8 8 2 5 0 2
9 5 6 3 0 3 9 0 8 7 1 8 2 7 1 3
6 6 6 4 8 9 7 9 8 3 3 0 8 3 1 8
9 7 1 2 6 0 0 7 6 3 1 8 5 5 7 6
0 6 8 6 7 9 0 3 4 8 9 7 8 3 8 9
6 3 0 1 0 3 5 9 5 8 0 3 1 7 8 8
8 7 4 5 7 9 1 7 2 0 7 8 3 3 0 4
3 3 9 0 1 8 2 9 0 0 9 0 1 3 4 5
3 6 7 0 0 7 9 7 4 0 4 0 4 1 6 0
```

- ☐ 543704
- ☐ 571053
- ☐ 281060
- ☐ 948773
- ☐ 759646
- ☐ 466095
- ☐ 173018
- ☐ 339018
- ☐ 008833
- ☐ 833664
- ☐ 361648
- ☐ 029008
- ☐ 985657
- ☐ 630103
- ☐ 563419
- ☐ 001187

Puzzle 11

```
9 4 3 4 0 7 8 5 5 1 9 2 6 3 7 9
1 6 4 5 4 7 4 6 6 1 7 3 2 2 8 3
8 9 6 4 4 7 3 0 3 8 0 4 0 6 6 7
7 1 4 7 7 1 5 1 3 4 0 5 5 7 4 0
8 6 0 7 4 4 6 0 8 8 8 4 8 8 9 9
9 5 3 4 5 0 4 3 2 0 4 9 5 7 5 7
7 5 6 0 3 6 1 6 9 2 5 9 5 0 6 7
8 5 6 9 3 9 9 2 8 3 1 0 0 2 4 1
5 3 0 9 1 2 3 6 8 3 2 1 8 2 3 5
4 1 9 2 9 8 0 6 4 8 4 5 5 7 9 4
7 1 0 8 9 0 0 6 9 0 1 2 6 8 6 6
5 3 0 1 3 0 8 7 9 3 5 5 8 9 2 9
5 5 9 0 3 9 4 5 4 4 5 5 5 2 8 2
2 7 9 4 8 7 6 7 4 4 9 0 8 1 4 2
7 4 7 6 0 3 4 5 1 1 5 0 0 4 9 5
3 2 5 0 2 4 2 7 0 7 5 0 4 6 5 0
5 7 6 9 9 0 1 7 2 0 8 4 4 7 2 6
7 1 2 7 1 2 9 0 1 1 0 9 1 3 2 7
```

- ☐ 563382
- ☐ 632190
- ☐ 904774
- ☐ 346403
- ☐ 099675
- ☐ 361692
- ☐ 155188
- ☐ 400511
- ☐ 710890
- ☐ 785475
- ☐ 242052
- ☐ 962849
- ☐ 513405
- ☐ 988488
- ☐ 192980
- ☐ 590394

Puzzle 12

```
7 4 3 9 4 5 2 6 0 8 7 7 0 4 8 6
1 9 5 4 4 5 3 0 8 6 5 0 1 5 0 0
2 0 4 5 1 6 2 9 6 8 8 5 9 3 4 7
8 2 7 0 5 2 1 3 3 7 3 1 4 4 9 3
4 9 5 3 3 2 8 7 7 7 4 0 1 4 6 8
6 5 5 4 1 5 6 3 9 6 3 2 8 8 7 6
2 5 0 2 5 6 7 8 4 4 7 9 3 0 4 8
9 3 3 7 8 3 7 3 6 8 9 8 5 5 7 9
4 5 2 5 6 9 8 1 4 9 5 3 5 1 4 4
3 6 6 4 1 8 1 9 0 2 8 3 4 2 5 5
3 3 8 1 4 1 4 7 3 3 2 9 4 4 4 6
0 3 8 2 0 0 6 3 6 4 0 6 5 9 2 6
0 0 2 0 6 6 5 6 0 4 8 3 6 2 4 7
4 9 3 3 2 6 2 6 9 8 6 2 2 7 9 0
3 8 3 4 6 6 2 1 2 5 6 4 6 0 4 3
8 5 6 2 0 9 6 5 2 9 4 7 5 0 2 0
6 7 5 8 0 0 1 6 2 1 9 0 6 8 0 3
3 1 2 5 5 3 9 9 4 8 4 7 1 8 0 5
```

- ☐ 554156
- ☐ 688308
- ☐ 212601
- ☐ 953328
- ☐ 046497
- ☐ 886230
- ☐ 854646
- ☐ 355445
- ☐ 124927
- ☐ 312507
- ☐ 937393
- ☐ 843086
- ☐ 967474
- ☐ 596611
- ☐ 508443
- ☐ 834003

Puzzle 13

```
9 2 6 0 9 1 4 1 5 0 0 0 4 7 7 5
4 9 3 6 8 4 7 1 2 8 1 6 3 3 7 3
4 3 0 0 5 6 3 0 0 3 8 5 3 1 4 3
7 1 6 9 5 3 7 6 3 1 9 4 6 7 1 4
7 4 4 9 0 5 4 3 9 5 7 4 2 3 5 6
4 5 4 2 5 9 6 0 3 4 1 4 1 8 9 6
3 7 6 3 4 5 6 1 4 3 8 8 2 1 0 2
7 5 5 6 7 6 6 8 3 7 6 3 9 2 5 0
3 0 8 1 8 7 6 9 3 1 9 8 9 4 7 0
4 2 8 3 0 6 9 2 6 8 2 8 8 1 2 0
6 9 0 3 7 2 6 0 6 0 8 0 8 5 7 1
2 5 5 9 9 4 2 3 7 9 1 4 0 1 5 6
4 6 7 7 3 6 4 8 7 5 6 4 1 5 3 1
6 3 0 4 4 2 8 7 5 0 0 3 5 6 9 8
1 4 7 1 0 1 3 2 4 2 4 4 7 3 8 8
2 8 4 1 9 4 1 9 2 7 1 6 6 3 8 3
6 0 5 5 9 3 6 1 7 6 2 7 7 9 5 0
9 9 2 0 7 6 3 4 7 3 4 1 8 9 8 1
```

- [ ] 749891
- [ ] 613842
- [ ] 724577
- [ ] 114793
- [ ] 359567
- [ ] 301892
- [ ] 384963
- [ ] 609923
- [ ] 668663
- [ ] 608085
- [ ] 703518
- [ ] 415000
- [ ] 188286
- [ ] 114759
- [ ] 314575
- [ ] 413583

Puzzle 14

```
2 5 1 5 7 1 0 4 6 2 1 5 3 7 0 5
1 9 5 8 0 7 0 5 2 7 3 7 5 4 6 4
8 5 5 3 8 6 2 0 1 1 2 3 8 1 2 7
4 1 8 8 5 5 9 0 8 2 8 1 9 0 6 7
3 7 3 6 1 8 1 5 8 2 5 6 2 7 4 1
0 4 4 7 5 3 4 1 9 1 5 7 7 3 7 8
2 2 0 8 0 0 5 6 0 5 7 4 1 6 3 2
6 1 3 4 3 4 2 2 2 1 0 2 8 5 6 9
4 6 1 1 1 2 2 8 0 0 9 7 2 2 9 5
3 3 4 4 4 8 0 3 8 1 7 1 6 1 9 8
2 4 0 7 8 9 6 1 8 7 8 3 3 2 7 3
5 6 4 0 6 6 7 5 9 2 2 2 8 6 4 1
1 5 4 8 0 7 0 6 9 5 5 6 2 7 3 7
5 2 0 4 4 9 2 6 1 0 4 5 1 1 2 5
1 6 6 1 4 2 6 0 0 7 2 1 8 6 5 8
2 8 3 1 7 9 7 7 9 3 5 6 0 7 1 7
7 8 8 6 3 1 9 3 7 2 0 9 9 5 8 9
3 5 8 5 4 7 4 7 1 4 9 9 0 0 0 4
```

- ☐ 851101
- ☐ 245287
- ☐ 900034
- ☐ 855709
- ☐ 276746
- ☐ 974325
- ☐ 122243
- ☐ 362817
- ☐ 836044
- ☐ 149761
- ☐ 729853
- ☐ 742713
- ☐ 684112
- ☐ 637462
- ☐ 365212
- ☐ 070527

Puzzle 15

```
5 6 3 5 5 8 9 4 0 2 8 2 1 0 2 7
4 4 9 7 6 0 1 7 2 7 0 3 7 2 7 2
5 2 5 5 9 7 5 9 9 6 8 2 1 2 5 8
7 8 6 2 8 1 1 4 5 0 8 9 0 3 4 2
4 4 8 4 4 8 5 0 6 8 0 8 6 6 4 9
1 4 3 4 8 2 1 7 3 8 1 3 0 0 2 0
7 3 9 5 8 5 1 0 2 0 6 9 3 5 1 6
8 5 8 0 6 3 3 3 3 3 2 0 4 6 8 6
7 8 0 6 2 9 2 5 7 1 5 2 6 9 7 6
9 0 7 6 6 8 7 3 2 2 9 2 7 5 1 7
1 1 7 5 4 1 1 3 8 7 2 3 7 4 9 9
1 2 9 5 7 0 3 0 6 9 0 9 4 8 8 0
6 6 4 2 9 1 0 7 8 6 9 8 0 9 2 2
0 3 9 9 3 8 7 4 8 4 4 8 0 3 5 4
3 8 5 5 5 5 2 3 5 4 9 0 0 0 3 2
9 9 5 2 2 0 5 4 7 0 7 8 0 4 7 6
1 6 8 3 6 9 1 1 4 5 0 9 8 1 5 3
9 2 9 2 3 0 8 8 1 3 4 0 6 5 6 0
```

- ☐ 995220
- ☐ 633009
- ☐ 808893
- ☐ 096030
- ☐ 563237
- ☐ 974626
- ☐ 306119
- ☐ 022360
- ☐ 545242
- ☐ 307271
- ☐ 801263
- ☐ 871982
- ☐ 466379
- ☐ 635589
- ☐ 445066
- ☐ 555523

Puzzle 16

```
2 5 4 1 2 8 8 6 2 4 9 7 4 0 6 5
8 6 0 9 9 6 0 3 4 2 2 5 8 1 7 9
5 0 3 9 7 4 8 7 2 5 1 6 1 0 2 3
9 7 7 8 7 0 5 2 7 9 8 0 3 6 4 6
5 3 0 5 3 1 3 9 2 0 2 1 5 5 4 4
5 6 6 5 6 1 8 6 1 9 2 6 7 5 2 8
4 2 3 6 2 5 9 9 4 2 3 6 1 9 6 0
0 8 2 6 4 2 7 2 8 8 0 8 5 7 8 7
8 3 3 7 5 9 7 1 5 3 0 5 5 3 4 2
6 2 4 7 0 6 6 8 4 4 1 5 3 3 0 2
4 8 0 9 2 3 4 4 4 7 8 7 8 4 9 1
1 7 5 3 8 5 1 6 8 0 1 5 5 0 0 4
3 8 2 7 7 7 4 0 5 3 7 5 4 2 7 1
8 9 4 3 7 1 3 5 2 4 9 0 9 7 9 8
4 4 6 5 9 2 2 4 2 3 8 0 6 7 6 8
4 5 1 2 4 6 7 0 6 1 5 1 0 4 3 9
3 7 6 7 8 1 4 3 3 4 4 8 0 7 2 3
4 5 6 8 9 3 3 2 0 4 0 1 7 3 0 6
```

- ☐ 548547
- ☐ 127242
- ☐ 403706
- ☐ 472341
- ☐ 595540
- ☐ 423619
- ☐ 704872
- ☐ 090796
- ☐ 566779
- ☐ 655012
- ☐ 238263
- ☐ 963571
- ☐ 372969
- ☐ 288085
- ☐ 234052
- ☐ 736245

Puzzle 17

```
6 9 6 4 0 2 6 2 5 8 3 7 9 5 4 5
4 6 1 9 8 9 1 7 7 9 5 3 7 3 7 7
7 4 8 0 3 8 9 8 0 9 3 5 5 8 1 7
2 1 1 0 2 7 3 2 7 2 4 9 7 3 1 0
1 5 3 0 3 8 1 5 7 6 7 4 2 3 1 9
7 5 0 3 6 4 8 0 9 4 2 2 2 4 0 6
5 8 1 3 8 9 3 9 2 5 9 7 2 6 9 9
4 7 6 3 6 5 3 2 6 6 3 2 8 7 0 4
4 5 9 4 7 7 7 9 2 9 3 7 8 6 6 6
4 8 2 2 5 0 9 4 8 2 5 0 5 8 0 5
4 4 6 0 6 4 6 2 5 8 5 2 8 7 0 9
0 7 6 8 2 1 6 5 2 0 9 8 4 6 0 6
1 3 1 3 6 7 5 7 4 6 4 8 5 5 7 9
6 8 6 3 0 3 2 8 1 1 0 7 6 0 1 9
4 9 4 7 4 5 7 1 6 9 4 0 0 0 2 7
4 3 9 6 4 1 4 4 5 4 7 6 6 2 4 0
9 9 9 1 5 9 6 3 3 2 7 1 1 6 3 9
2 6 8 7 2 5 8 0 4 8 9 8 2 9 7 2
```

☐ 992749   ☐ 867562   ☐ 490003   ☐ 752492

☐ 083236   ☐ 272076   ☐ 714075   ☐ 926285

☐ 000496   ☐ 217000   ☐ 643383   ☐ 947457

☐ 368684   ☐ 266745   ☐ 788207   ☐ 975722

Puzzle 18

```
3 2 7 3 9 4 1 9 5 3 7 9 0 0 6 8
7 7 8 4 2 3 1 9 8 9 9 6 1 6 4 6
2 7 7 8 7 8 6 6 1 1 0 8 9 3 4 2
9 5 0 8 5 6 7 5 9 6 3 8 7 4 0 7
9 6 0 0 4 2 2 1 2 0 8 1 7 1 0 9
6 5 8 9 8 0 1 4 8 2 9 9 3 9 5 7
6 9 8 5 8 8 5 0 9 3 0 7 1 9 7 7
0 0 1 1 2 1 1 0 4 9 7 4 1 8 6 0
1 4 9 5 8 0 9 7 3 3 5 1 9 4 4 0
4 3 5 6 6 3 1 1 6 6 7 0 6 2 7 8
1 0 4 1 5 9 3 9 3 1 7 3 2 9 9 1
8 5 5 4 4 0 6 2 6 4 9 8 4 5 9 3
7 8 6 4 9 2 1 6 0 8 8 1 5 6 9 1
2 5 8 2 3 2 2 2 1 8 6 1 4 1 9 5
4 1 6 8 7 1 5 3 6 8 4 3 5 5 0 6
0 3 9 8 8 1 1 2 0 1 3 1 7 3 1 4
0 7 3 6 5 7 5 8 6 6 4 4 9 0 5 2
3 7 7 4 0 4 6 1 4 3 6 5 2 4 8 5
```

- [ ] 711220
- [ ] 140071
- [ ] 916023
- [ ] 669927
- [ ] 966183
- [ ] 102118
- [ ] 073686
- [ ] 396865
- [ ] 142230
- [ ] 187240
- [ ] 215191
- [ ] 043058
- [ ] 335194
- [ ] 148023
- [ ] 778786
- [ ] 842319

Puzzle 19

```
7 8 3 5 9 6 8 4 0 5 9 6 1 0 5 0
7 0 1 3 0 9 6 3 9 8 7 8 3 7 2 7
4 3 7 6 4 8 0 2 0 5 5 5 6 5 3 2
6 7 7 5 0 7 0 8 1 9 7 3 2 5 6 9
0 5 8 7 1 5 2 7 3 7 4 7 7 4 6 7
7 4 0 4 2 9 5 0 8 8 6 1 0 6 1 9
1 4 5 3 1 3 0 8 0 2 7 9 6 4 3 0
7 1 0 4 3 3 7 7 2 9 4 9 9 9 6 0
7 8 9 9 0 8 3 5 8 7 7 7 1 7 5 2
5 9 0 1 6 3 4 3 1 0 9 4 6 3 0 8
5 9 1 5 2 6 5 6 3 7 6 4 8 2 5 8
3 0 5 1 8 5 6 9 7 8 3 3 6 1 2 3
3 3 0 9 8 7 5 3 7 1 8 7 2 7 4 2
3 6 9 6 9 3 6 5 6 8 8 1 3 2 9 8
6 1 2 4 0 8 6 3 1 9 5 1 4 6 9 7
8 4 0 8 4 8 9 7 8 1 3 4 3 5 0 4
6 2 0 6 2 7 4 5 7 8 0 4 6 1 0 2
7 5 7 3 3 6 6 0 1 1 5 3 9 3 6 9
```

☐ 907806   ☐ 942505   ☐ 405961   ☐ 194347

☐ 507081   ☐ 699949   ☐ 754726   ☐ 936903

☐ 240988   ☐ 285506   ☐ 415913   ☐ 960726

☐ 537187   ☐ 265637   ☐ 387965   ☐ 853719

Puzzle 20

```
1 3 1 3 7 7 7 5 0 0 3 6 7 2 8 3
1 8 8 6 9 1 6 8 4 2 9 0 2 4 5 1
3 5 6 5 9 6 2 1 9 5 8 6 1 6 8 6
6 3 1 0 3 1 9 6 5 8 7 7 2 0 1 7
2 7 5 4 6 1 7 0 1 2 3 1 4 9 3 1
2 2 9 0 6 9 8 0 6 5 1 9 7 5 7 5
8 3 3 8 6 4 7 0 6 1 9 2 9 1 1 7
7 5 0 3 7 1 4 9 0 1 2 8 0 0 7 9
7 9 7 7 1 8 2 6 1 7 2 4 8 2 4 7
1 9 1 4 9 1 8 6 3 4 4 1 8 3 7 3
4 5 3 3 2 2 3 0 3 0 0 0 1 0 8 3
4 0 8 5 5 1 3 2 3 6 6 7 7 2 3 4
7 5 2 2 7 8 8 9 8 4 2 9 8 7 7 5
9 9 2 4 0 9 4 5 2 0 9 0 4 6 3 1
3 4 6 3 3 5 3 0 7 6 6 9 5 0 0 9
1 5 3 1 4 8 9 0 5 3 8 5 1 6 6 6
7 0 0 7 3 7 7 9 7 1 9 0 2 1 4 2
3 8 7 6 8 7 0 5 5 2 1 9 3 3 6 8
```

- [ ] 896092
- [ ] 153148
- [ ] 603738
- [ ] 987880
- [ ] 569130
- [ ] 025490
- [ ] 005777
- [ ] 291378
- [ ] 516217
- [ ] 636205
- [ ] 797190
- [ ] 138226
- [ ] 397441
- [ ] 140786
- [ ] 581371
- [ ] 061211

Puzzle 21

```
0 3 6 7 1 6 9 4 6 1 8 4 7 6 5 8
6 5 7 3 9 0 5 1 6 4 0 0 8 2 9 3
6 7 0 3 3 1 0 5 1 8 5 4 6 0 8 7
8 0 9 6 2 2 5 7 9 2 8 1 7 9 5 0
6 6 3 5 6 7 2 4 4 5 9 5 6 0 5
4 7 4 7 0 6 2 5 2 0 3 3 0 5 6 4
5 1 9 8 5 9 8 6 1 5 6 7 8 2 7 8
7 1 0 0 8 3 4 3 0 2 4 8 5 7 9 9
2 9 5 6 1 4 8 1 2 6 8 4 5 4 6 7
3 7 0 4 0 8 0 8 4 1 8 3 1 4 3 8
1 7 1 1 9 5 1 4 8 9 2 6 1 0 9 1
1 7 6 8 5 1 4 0 4 0 6 0 8 4 0 9
3 0 9 9 4 7 4 4 6 2 8 6 8 9 7 6
3 3 7 6 3 2 5 8 1 5 8 9 3 4 1 3
9 2 7 1 0 4 5 1 5 3 7 6 0 6 7 8
6 9 1 1 9 6 6 9 7 8 0 4 4 9 9 7
8 4 7 3 5 3 4 2 3 9 1 4 6 2 1 7
3 1 1 7 0 1 9 0 4 3 4 0 8 4 1 3
```

- ☐ 144527
- ☐ 048136
- ☐ 986860
- ☐ 460875
- ☐ 573495
- ☐ 309947
- ☐ 945860
- ☐ 911253
- ☐ 657390
- ☐ 291840
- ☐ 914858
- ☐ 096527
- ☐ 950522
- ☐ 158934
- ☐ 468890
- ☐ 748164

Puzzle 22

```
8 9 2 5 0 2 5 4 8 2 8 8 5 8 9 4
3 6 7 8 5 6 8 6 0 4 2 0 4 9 2 6
8 4 2 3 0 6 5 5 6 0 4 9 6 2 9 8
7 0 7 5 5 1 3 0 1 9 0 7 5 6 5 1
0 2 9 9 5 1 5 9 9 5 3 5 6 9 1 1
6 0 0 9 0 1 0 7 0 4 4 1 4 4 4 0
2 8 7 2 7 7 2 4 0 4 2 3 6 3 1 4
7 1 0 1 5 4 8 6 1 1 7 0 9 7 7 1
5 4 9 6 9 7 9 4 0 2 7 9 1 8 5 6
1 8 7 7 9 9 4 8 6 8 8 7 5 3 8 1
1 4 3 3 0 9 8 3 4 9 6 0 4 5 8 8
3 4 4 3 8 4 3 2 6 0 1 3 6 2 5 8
4 7 9 4 8 5 1 3 8 9 8 9 7 8 8 1
4 2 8 6 3 6 1 0 7 2 0 8 2 4 4 8
6 1 1 0 3 5 6 5 9 1 5 7 8 4 3 4
5 9 7 5 7 0 1 5 2 5 0 9 7 4 0 6
5 8 5 6 9 0 3 6 1 4 6 0 2 4 9 0
1 2 2 8 1 6 1 2 9 1 1 2 1 3 0 1
```

- [ ] 140153
- [ ] 892694
- [ ] 513097
- [ ] 216733
- [ ] 070973
- [ ] 733880
- [ ] 208148
- [ ] 892502
- [ ] 295282
- [ ] 948880
- [ ] 069438
- [ ] 641630
- [ ] 367856
- [ ] 260863
- [ ] 799456
- [ ] 964656

Puzzle 23

```
9 9 7 8 7 5 5 0 5 5 5 3 6 2 9 4 7
0 5 8 8 7 1 8 6 6 7 1 6 5 0 0 9
9 4 3 8 5 9 8 3 1 1 7 7 9 3 3 7
6 6 9 7 6 8 9 9 2 6 5 7 4 3 1 0
0 4 2 2 1 9 7 8 4 1 7 2 5 0 4 9
7 5 8 5 0 2 2 0 2 4 0 3 4 3 6 5
9 7 0 3 9 7 2 4 5 3 1 4 0 6 5 5
5 2 7 7 5 1 0 5 9 5 8 6 5 1 9 1
4 8 5 5 4 6 6 1 4 6 7 0 9 2 1 0
5 0 0 8 0 0 8 4 7 3 2 6 6 7 0 1
4 2 2 6 2 9 7 4 3 2 6 5 5 9 4 6
0 6 1 4 3 6 4 9 5 4 9 8 2 2 6 1
1 2 4 5 8 4 7 7 1 3 7 2 2 0 4 9
2 6 2 2 1 0 7 9 0 8 9 3 5 2 1 8
8 0 9 9 1 6 3 3 2 8 3 7 0 9 4 6
8 4 6 9 5 3 7 0 1 8 8 1 2 1 5 7
6 7 2 0 5 0 5 2 7 2 2 0 5 2 8 5
1 7 1 2 8 2 3 3 4 9 1 5 3 5 6 6
```

- ☐ 684539
- ☐ 186671
- ☐ 875505
- ☐ 796278
- ☐ 706909
- ☐ 104545
- ☐ 210790
- ☐ 537018
- ☐ 343042
- ☐ 785870
- ☐ 465910
- ☐ 207017
- ☐ 978417
- ☐ 050527
- ☐ 122543
- ☐ 214296

Puzzle 24

```
8 1 6 1 7 7 6 7 7 4 6 9 9 9 8 7 9
7 0 5 7 8 2 7 3 0 2 3 0 1 7 8 9
2 4 7 5 3 2 7 4 0 9 9 0 7 0 2 0
9 7 9 3 2 0 5 2 7 0 9 9 2 7 9 4
7 2 4 7 9 7 4 8 3 4 3 5 8 5 4 8
3 2 8 7 2 9 9 5 0 7 4 6 0 6 8 8
9 1 8 4 7 9 4 5 6 6 5 4 7 0 4 9
9 3 0 9 2 1 1 4 5 6 8 6 6 8 3 8
0 9 2 2 4 8 1 9 6 5 7 4 3 7 2 9
4 1 9 0 8 3 0 9 4 6 0 4 9 0 3 1
1 6 9 1 1 5 8 7 2 9 9 8 5 0 7 6
5 0 7 7 0 1 9 9 1 7 7 6 0 5 0 2
9 3 9 1 1 2 6 8 4 8 8 0 3 9 2 8
5 7 9 3 2 9 6 9 1 9 7 7 7 0 6 4
1 5 3 4 0 3 9 1 6 7 4 3 9 0 8 8
7 3 7 8 1 0 1 1 0 1 0 2 8 5 2 1
7 2 0 8 2 2 2 5 7 3 1 4 7 6 4 6
4 8 7 3 2 8 4 2 4 5 7 8 2 4 5 4
```

- ☐ 870059
- ☐ 842290
- ☐ 787291
- ☐ 729194
- ☐ 055597
- ☐ 904766
- ☐ 287507
- ☐ 971991
- ☐ 159517
- ☐ 177472
- ☐ 689927
- ☐ 082225
- ☐ 116961
- ☐ 771618
- ☐ 843171
- ☐ 994466

Puzzle 25

```
6 0 0 2 5 5 9 7 0 8 1 7 2 3 3 3
6 1 1 4 9 5 0 9 1 1 1 7 4 3 7 4
9 4 5 5 7 9 5 7 6 7 8 2 4 2 8 4
5 6 4 4 5 9 5 7 3 8 4 1 6 4 1 3
1 5 3 3 2 9 8 0 4 5 4 3 8 5 8 5
0 0 0 6 2 1 3 2 0 1 5 1 3 1 2 7
7 8 4 5 7 8 9 7 0 6 3 6 7 7 2 7
4 5 1 8 5 2 0 2 8 8 9 3 9 2 7 8
3 2 8 8 0 0 3 8 4 8 2 1 6 8 4 5
4 6 4 1 9 5 8 5 9 3 0 8 7 4 8 2
3 8 9 6 4 1 3 9 3 5 1 9 6 7 8 0
4 1 5 3 6 5 4 0 9 8 8 9 4 5 4 0
6 9 9 9 0 5 1 7 9 1 2 8 9 3 3 9
4 6 6 4 5 4 1 4 4 5 5 0 4 0 6 5
8 5 3 6 3 6 0 9 9 6 1 6 5 8 3 0
8 0 7 6 1 2 5 4 7 7 1 5 1 3 0 4
3 9 0 2 0 7 0 8 2 1 3 3 3 4 9 1
4 9 6 3 4 2 0 1 2 9 2 1 0 3 0 9
```

- ☐ 055055
- ☐ 182511
- ☐ 309385
- ☐ 185388
- ☐ 018181
- ☐ 772131
- ☐ 958272
- ☐ 967649
- ☐ 644328
- ☐ 029354
- ☐ 791566
- ☐ 728475
- ☐ 403963
- ☐ 781822
- ☐ 319432
- ☐ 082133

Puzzle 26

```
9 2 1 1 2 2 8 8 3 6 3 1 9 5 1 3
6 0 2 1 0 4 4 1 3 7 5 4 7 3 0 5
6 5 1 6 8 3 2 4 1 4 9 3 2 5 3 9
8 3 9 0 4 1 0 3 0 5 8 6 5 6 3 4
2 7 5 0 8 5 5 5 5 4 5 6 7 1 6 5
7 4 1 3 8 5 8 0 2 8 2 6 4 3 4 0
5 1 3 5 4 1 6 7 2 8 5 8 1 5 4 0
8 9 3 4 7 6 1 2 0 0 4 5 5 7 4 2
6 0 3 1 2 4 9 8 6 0 9 4 9 8 6 3
9 0 8 1 1 3 8 4 5 9 8 1 4 9 8 4
0 1 3 4 4 4 8 0 0 0 5 9 0 8 7 9
8 4 8 1 9 6 7 5 1 9 1 4 8 4 5 9
0 4 6 6 0 3 9 0 4 6 5 1 2 7 5 9
2 2 6 0 2 6 9 1 1 3 3 4 8 5 6 1
2 9 5 5 8 6 2 2 3 8 6 0 1 7 7 2
7 9 5 2 6 7 8 4 8 2 5 2 1 0 5 7
1 9 6 2 6 7 5 3 6 6 9 2 9 6 9 7
5 3 5 1 4 4 5 7 7 4 7 1 4 9 3 2
```

- ☐ 748848
- ☐ 145904
- ☐ 221129
- ☐ 615513
- ☐ 039046
- ☐ 762597
- ☐ 149325
- ☐ 440428
- ☐ 940828
- ☐ 626753
- ☐ 235855
- ☐ 208096
- ☐ 598149
- ☐ 861988
- ☐ 260269
- ☐ 875567

Puzzle 27

```
0 6 8 8 1 2 1 5 2 3 4 1 5 6 0 3
0 4 9 8 7 3 2 7 0 7 2 4 3 2 2 5
1 3 3 1 0 6 8 0 0 0 3 1 1 4 3 7
3 8 7 0 2 1 7 9 2 3 4 9 9 7 7 1
3 4 8 9 6 3 5 0 4 4 6 6 3 5 0 0
7 3 9 1 1 4 8 4 9 3 8 1 3 8 4 6
4 9 4 6 4 8 6 2 0 4 2 1 7 5 9 7
8 8 5 8 5 7 2 6 6 7 2 5 1 3 4 2
7 2 0 9 7 4 8 8 2 2 3 7 8 5 4 3
5 9 5 9 6 4 2 2 3 8 8 3 8 3 2 4
1 1 3 2 7 7 9 6 8 5 5 2 4 1 8 5
2 2 0 5 7 5 7 7 4 8 4 0 3 8 6 7
6 7 9 2 9 6 6 5 1 7 7 3 1 3 8 3
9 1 2 0 8 1 6 6 8 6 3 6 2 9 6 0
4 4 2 4 1 5 4 5 0 2 9 7 9 9 2 1
1 6 0 5 1 8 6 8 6 6 4 4 9 2 4 2
9 9 0 1 2 0 6 9 2 2 0 0 6 2 1 4
0 9 7 8 9 0 2 2 4 3 9 1 7 5 1 9
```

- [ ] 255869
- [ ] 220248
- [ ] 704510
- [ ] 617947
- [ ] 496542
- [ ] 309220
- [ ] 382628
- [ ] 716307
- [ ] 430646
- [ ] 004635
- [ ] 490767
- [ ] 085165
- [ ] 664492
- [ ] 273938
- [ ] 049442
- [ ] 512694

Puzzle 28

```
6 3 8 8 9 6 8 7 1 0 9 2 6 9 7 7
1 3 1 5 7 4 5 4 6 1 8 1 6 9 9 6
7 1 7 7 9 6 9 0 8 7 6 7 7 3 0 0
2 9 3 6 6 6 8 5 1 2 6 5 5 9 6 2
3 8 3 9 5 4 8 2 2 1 7 7 3 0 4 3
0 1 4 8 3 3 0 0 3 3 6 1 7 2 0 9
2 5 0 1 6 3 6 4 0 2 4 5 7 1 8 6
7 6 6 9 4 8 7 6 4 9 0 7 8 2 8 6
9 1 9 0 4 4 7 9 5 2 5 5 0 8 7 0
5 5 7 7 6 9 2 1 2 8 7 0 8 0 0 3
2 8 1 7 5 6 2 4 7 8 3 7 7 9 1 4
5 2 6 0 4 7 4 2 6 0 1 1 5 3 4 8
4 6 4 4 5 4 6 4 1 6 0 2 4 5 8 6
3 7 8 0 6 8 3 8 5 1 4 2 5 6 0 4
6 5 6 8 9 9 1 6 5 8 7 1 5 0 5 3
5 7 6 6 3 8 9 7 7 2 1 4 0 1 0 5
8 2 7 7 3 8 3 9 0 2 9 4 7 5 1 8
8 8 7 6 1 4 8 8 9 3 0 7 6 4 1 7
```

☐ 463569	☐ 996468	☐ 768424	☐ 665831
☐ 390821	☐ 808754	☐ 786988	☐ 007178
☐ 967755	☐ 698190	☐ 721401	☐ 838865
☐ 505178	☐ 772440	☐ 547513	☐ 675377

Puzzle 29

```
1 7 9 0 8 8 9 9 7 4 4 8 1 6 7 7
5 0 4 8 9 0 4 2 5 1 3 7 6 7 4 0
4 5 4 8 5 8 0 5 7 9 3 5 4 0 7 2
3 1 4 3 6 5 1 6 6 4 6 9 0 3 5 5
0 7 0 1 7 6 0 9 7 6 6 9 6 3 6 0
1 3 7 0 0 0 7 7 7 1 4 2 6 5 1 2
6 3 6 3 1 4 2 7 5 3 5 7 8 5 6 1
6 4 1 4 5 9 6 6 5 8 3 7 1 9 5 0
5 3 7 3 9 8 0 7 4 7 3 3 7 5 8 1
3 0 2 5 6 2 2 7 4 1 1 9 9 3 2 9
5 4 7 9 8 3 2 4 7 6 1 2 8 7 9 7
9 1 6 1 7 9 5 8 1 8 1 2 7 0 3 5
3 9 7 2 3 4 2 7 3 1 2 6 5 1 0 9
7 3 4 4 4 1 2 3 9 7 2 9 0 8 6 6
6 5 4 8 8 8 6 5 8 3 9 7 7 5 3 9
9 0 4 9 4 2 6 6 5 9 1 2 4 1 3 7
3 2 7 6 8 8 8 8 5 2 0 4 0 3 2 1
1 0 8 1 1 1 3 8 5 5 0 5 7 9 2 5
```

☐ 701851	☐ 622741	☐ 704539	☐ 746654
☐ 484378	☐ 917385	☐ 767796	☐ 603928
☐ 641459	☐ 779385	☐ 543016	☐ 798324
☐ 962293	☐ 447998	☐ 040258	☐ 252266

Puzzle 30

```
8 7 4 6 6 6 0 0 7 4 7 2 1 1 2 5
3 3 7 5 1 3 6 7 8 0 6 9 6 7 8 8
6 6 3 2 5 2 6 8 3 4 3 6 7 6 7 4
4 7 5 2 9 4 2 2 6 8 4 4 7 9 2 6
5 6 5 4 6 9 5 4 3 4 9 5 8 9 4 4
4 7 8 2 2 4 8 1 5 7 6 0 7 8 3 2
4 9 6 9 3 5 7 8 1 7 1 2 5 7 8 3
1 6 8 1 4 9 5 7 0 4 3 3 3 4 2 0
1 9 8 4 7 0 8 8 4 7 2 6 8 2 2 3
3 4 2 6 8 6 4 6 9 2 0 2 3 6 7 2
5 8 8 8 5 2 1 9 0 6 6 3 6 4 9 5
2 4 7 7 4 7 7 6 4 9 1 0 9 8 5 8
1 0 2 2 0 4 4 1 9 4 5 1 1 1 2 9
2 7 0 0 9 0 5 2 0 9 5 8 5 2 5 3
2 1 3 0 6 8 0 3 0 1 6 2 6 3 1 3
8 0 6 1 7 6 7 2 9 3 4 2 2 1 9 9
3 4 0 0 7 4 3 1 0 2 2 9 0 3 3 5
1 9 9 7 3 9 8 1 1 7 1 6 7 4 0 9
```

- ☐ 070391
- ☐ 291468
- ☐ 224925
- ☐ 418786
- ☐ 243927
- ☐ 125888
- ☐ 764910
- ☐ 086031
- ☐ 862241
- ☐ 769987
- ☐ 450983
- ☐ 521228
- ☐ 962347
- ☐ 274700
- ☐ 367679
- ☐ 714857

Puzzle 31

```
5 6 7 4 4 1 9 1 1 1 5 7 6 4 5 3
3 2 9 1 9 4 2 5 8 3 6 9 5 5 6 7
3 5 4 2 3 9 3 1 2 5 6 2 7 6 7 1
5 8 5 0 5 3 3 6 4 4 8 7 7 5 6 9
3 2 1 7 6 6 0 8 7 9 8 6 5 7 3 0
6 2 7 6 5 2 7 7 9 0 4 3 9 1 4 9
8 9 7 7 2 5 7 9 1 5 1 8 7 6 1 7
3 4 1 4 2 0 8 4 6 7 1 8 3 3 0 4
9 2 3 1 4 6 7 6 7 4 1 2 1 9 4 0
6 0 5 3 5 3 1 7 6 3 0 4 3 9 8 5
0 6 5 6 3 4 9 6 6 5 6 1 9 5 5 4
9 4 6 4 1 4 9 3 1 4 2 5 6 5 1 0
0 4 3 8 8 0 4 8 9 4 6 9 8 0 7 6
1 0 0 2 7 8 8 8 3 8 8 5 3 6 1 9
9 9 7 7 0 4 3 8 7 0 5 7 4 3 9 7
7 6 8 9 5 4 6 4 6 0 7 1 2 3 3 7
5 0 5 0 2 1 3 0 3 4 1 8 7 7 6 5
5 8 5 5 2 9 8 4 3 6 7 5 1 4 5 9
```

- ☐ 772616
- ☐ 610886
- ☐ 402340
- ☐ 491923
- ☐ 494839
- ☐ 565241
- ☐ 964984
- ☐ 647866
- ☐ 764978
- ☐ 021303
- ☐ 191447
- ☐ 248373
- ☐ 524062
- ☐ 934036
- ☐ 104851
- ☐ 767265

Puzzle 32

```
9 4 5 6 7 1 3 2 3 7 0 4 9 9 4 9
8 2 7 9 2 5 7 3 1 3 5 2 5 1 4 8
8 9 4 9 0 0 6 5 2 6 0 8 0 6 6 2
4 0 5 9 4 8 9 5 0 7 2 7 8 4 0 8
2 9 1 9 4 6 5 3 1 1 8 2 3 0 0 3
0 5 3 6 8 8 3 8 1 1 4 1 2 1 2 9
3 0 6 4 9 6 5 5 8 1 4 0 1 2 2 5
0 4 1 4 2 3 0 8 3 2 8 4 0 0 0 1
1 3 1 4 7 0 4 3 1 4 1 7 3 3 1 1
3 7 2 8 1 5 3 0 0 4 4 3 1 4 5 5
4 6 5 7 0 1 7 2 8 9 1 8 1 9 2 0
8 2 8 5 2 9 1 9 4 3 7 9 7 8 7 4
9 6 8 1 1 7 6 1 1 3 6 5 6 0 9 5
5 5 7 8 2 7 2 1 1 8 7 2 1 7 3 1
5 8 9 2 7 0 9 8 3 7 0 6 0 1 5 8
9 2 5 8 1 2 2 0 4 2 8 2 3 8 0 0
9 1 3 6 3 1 4 3 5 0 6 7 1 9 3 5
6 1 6 1 7 1 1 8 2 6 4 6 3 1 3 6
```

- [ ] 804396
- [ ] 948950
- [ ] 204282
- [ ] 975746
- [ ] 711301
- [ ] 485814
- [ ] 916401
- [ ] 024889
- [ ] 525313
- [ ] 909504
- [ ] 353448
- [ ] 712781
- [ ] 072985
- [ ] 278824
- [ ] 631435
- [ ] 740127

Puzzle 33

```
8 7 1 0 7 4 6 3 6 8 5 2 7 3 6 0
0 5 6 0 1 7 6 0 2 0 4 1 8 4 3 1
8 0 3 5 1 1 1 4 1 6 7 1 8 9 2 8
5 5 4 0 6 4 3 2 7 2 3 1 4 3 5 4
5 4 1 1 7 6 8 3 0 2 0 9 3 4 8 5
3 7 4 7 1 4 5 1 3 0 6 7 0 5 6 8
6 4 5 0 9 9 4 8 4 7 8 4 8 0 1 8
6 5 7 9 7 1 6 4 2 4 5 8 5 8 5 5
2 0 4 9 6 8 2 6 4 0 4 0 2 2 7 4
0 8 1 1 0 6 0 2 0 2 6 7 6 3 1 3
3 4 4 0 4 6 4 4 4 9 8 4 9 1 3 8
5 8 9 9 2 1 3 8 6 3 2 4 1 7 8 2
1 3 5 3 3 1 2 9 8 2 1 5 6 6 6 5
5 4 4 4 2 5 0 0 7 5 4 7 6 8 7 9
9 1 0 4 1 7 0 5 2 8 4 8 2 8 7 0
2 5 7 9 8 2 8 8 2 8 5 8 5 6 6 1
7 9 7 4 7 7 3 1 4 2 4 1 6 2 2 2
3 4 3 2 7 1 6 5 3 9 0 9 3 3 2 0
```

- ☐ 417052
- ☐ 887021
- ☐ 676202
- ☐ 627464
- ☐ 882858
- ☐ 263900
- ☐ 601180
- ☐ 851791
- ☐ 457005
- ☐ 725863
- ☐ 440780
- ☐ 514384
- ☐ 773142
- ☐ 776831
- ☐ 035111
- ☐ 920470

Puzzle 34

```
2 9 7 9 9 6 1 7 0 4 7 9 3 9 8 4
4 1 9 7 0 3 1 0 7 0 9 3 4 9 6 6
3 8 3 1 6 9 9 5 4 7 4 1 2 7 0 7
0 0 7 1 6 1 0 6 8 0 4 6 6 0 5 2
7 8 1 1 9 9 9 3 3 8 2 5 9 2 7 9
5 9 4 9 1 8 2 3 1 3 7 5 2 7 8 2
0 8 2 6 8 7 7 1 7 5 8 4 4 7 7 3
1 3 3 0 4 8 5 0 8 5 6 6 4 5 8 7
6 6 2 7 7 1 9 7 7 0 3 1 4 5 2 2
3 1 0 3 1 5 6 7 2 7 1 0 5 6 7 3
9 2 8 6 6 9 9 1 7 9 2 1 5 9 3 9
4 7 3 0 9 1 3 4 4 7 1 3 5 6 7 8
7 6 7 1 0 1 3 5 3 8 0 2 5 6 9 6
6 2 0 8 5 0 5 3 3 9 9 7 7 2 1 4
3 6 1 7 1 1 8 9 8 9 8 0 4 6 0 8
5 1 2 2 1 5 1 0 4 1 5 3 0 4 5 4
8 1 5 1 1 8 7 7 9 0 9 2 9 8 5 0
9 8 9 9 2 4 0 8 2 5 7 7 4 0 7 6
```

- ☐ 161068
- ☐ 961383
- ☐ 093541
- ☐ 924445
- ☐ 419703
- ☐ 472787
- ☐ 655772
- ☐ 811518
- ☐ 960736
- ☐ 208370
- ☐ 723101
- ☐ 878750
- ☐ 613381
- ☐ 716997
- ☐ 471690
- ☐ 591101

Puzzle 35

```
0 1 6 2 5 9 3 2 7 9 8 8 2 4 4 8
2 0 7 1 2 2 9 8 7 1 4 8 6 0 4 5
0 1 3 1 6 8 1 1 9 6 2 7 3 6 4 7
6 1 0 5 2 6 7 5 8 3 5 4 2 4 3 9
0 8 3 5 4 8 3 0 2 4 4 2 8 6 7 6
1 8 7 5 7 8 0 2 7 5 7 6 6 6 5 5
4 5 6 6 7 7 8 2 2 7 2 7 9 7 7 5
7 8 5 2 8 5 3 8 8 2 4 9 4 4 2 7
4 5 6 7 6 4 9 8 4 0 5 1 7 4 2 1
2 4 0 5 3 5 9 0 8 5 4 0 6 1 0 4
4 6 9 8 0 5 7 1 0 2 0 6 7 8 0 0
9 5 5 4 5 3 1 1 8 1 7 7 4 2 3 5
7 8 3 7 8 5 0 5 2 5 3 9 8 4 6 7
9 1 3 6 7 1 6 6 9 3 0 5 3 4 2 1
8 4 7 2 4 8 4 9 6 2 0 8 4 2 6 7
0 8 6 1 6 1 3 7 4 5 0 9 9 0 4 5
4 7 5 8 7 7 0 2 2 7 3 8 8 5 7 3
4 1 6 4 3 7 6 6 8 0 6 8 4 6 6 4
```

- ☐ 001354
- ☐ 681196
- ☐ 275847
- ☐ 727228
- ☐ 785036
- ☐ 684918
- ☐ 854061
- ☐ 003548
- ☐ 029515
- ☐ 632869
- ☐ 564585
- ☐ 036264
- ☐ 354830
- ☐ 842547
- ☐ 359065
- ☐ 077857

Puzzle 36

```
3 3 8 0 7 8 1 9 6 7 7 3 4 7 9 4
0 9 3 4 5 0 4 9 8 6 0 6 0 2 2 1
6 7 2 6 0 2 2 3 9 1 6 7 4 8 0 2
4 9 3 6 9 6 8 0 9 2 3 2 6 5 7 6
3 8 5 6 7 8 2 7 0 1 5 6 9 5 2 1
5 1 5 8 9 8 4 5 7 8 1 2 6 2 6 8
9 5 7 0 8 6 9 2 0 9 7 8 0 0 0 6
2 7 9 4 6 7 1 5 5 0 7 1 3 6 4 3
1 5 6 5 8 3 4 7 4 8 5 0 0 0 9 9
4 0 3 4 5 0 0 0 1 0 3 2 3 8 8 7
1 6 4 3 7 5 0 1 4 8 2 8 0 6 1 7
7 8 2 6 5 7 0 7 4 0 6 6 0 4 7 3
5 8 0 4 5 7 0 9 2 5 2 3 9 2 7 5
2 7 6 8 8 5 5 9 8 2 1 0 2 8 1 5
6 8 5 9 4 9 3 1 5 3 3 8 8 1 5 6
6 4 4 9 3 7 8 0 3 6 1 5 6 0 0 5
7 7 8 2 5 8 9 7 2 2 5 2 3 0 3 3
3 2 1 0 4 1 4 5 8 8 4 1 6 9 7 5
```

- ☐ 012368
- ☐ 322700
- ☐ 377691
- ☐ 035564
- ☐ 608642
- ☐ 232908
- ☐ 920978
- ☐ 890555
- ☐ 385678
- ☐ 257141
- ☐ 255827
- ☐ 726022
- ☐ 517649
- ☐ 934504
- ☐ 685949
- ☐ 391674

Puzzle 37

```
6 9 3 2 3 9 3 1 4 0 6 1 8 6 7 4
6 8 3 2 8 9 0 1 9 4 0 0 9 1 8 7
3 0 9 8 9 9 5 6 0 3 4 4 7 4 2 9
9 4 9 7 4 7 4 4 7 4 0 1 4 8 2 2
8 2 0 3 7 1 6 6 2 7 1 9 7 3 7 5
0 2 0 7 7 4 3 0 4 5 4 4 9 0 0 6
7 1 6 1 1 5 9 9 0 9 5 0 0 8 1 8
4 0 6 4 0 3 9 2 9 0 9 8 9 9 5 8
4 1 0 4 3 0 1 5 7 1 2 3 3 1 8 1
5 0 3 2 2 8 5 8 5 4 5 3 9 2 8 5
8 8 1 0 4 1 5 3 4 4 6 7 8 5 9 9
0 0 7 6 2 0 7 9 3 4 4 9 2 1 3 5
4 7 0 5 5 4 4 3 0 1 8 6 3 6 3 1
3 8 0 7 8 4 1 9 1 3 4 8 9 1 3 7
6 2 1 1 1 0 7 2 6 3 9 5 2 1 5 4
8 3 5 2 4 6 3 6 2 3 9 3 8 5 8 5
3 0 0 2 8 9 9 8 3 7 4 7 5 9 3 4
7 3 7 2 7 0 2 9 2 7 5 8 3 4 7 3
```

☐ 859039 ☐ 599095 ☐ 899820 ☐ 980744

☐ 875803 ☐ 510722 ☐ 484652 ☐ 483089

☐ 160413 ☐ 289464 ☐ 067740 ☐ 544301

☐ 347702 ☐ 448131 ☐ 796853 ☐ 293289

Puzzle 38

```
5 5 2 9 1 5 1 7 9 7 4 1 7 2 1 0
7 4 6 3 5 1 9 2 1 0 2 0 1 4 2 5
6 8 3 4 2 8 4 8 3 6 2 1 8 0 5 0
5 3 8 3 0 7 0 3 6 0 2 0 4 7 6 6
5 2 5 0 8 4 7 6 2 2 2 0 4 7 3 5
9 9 7 5 5 4 2 4 5 7 7 8 7 7 1 9
9 8 2 8 7 8 3 9 1 3 6 0 1 1 9 8
9 0 5 7 9 1 4 6 5 5 5 1 8 4 3 1
3 1 4 8 0 9 3 2 3 7 0 2 0 8 7 0
4 4 9 1 2 6 1 2 3 3 9 9 9 0 8 7
0 0 5 5 2 1 4 9 5 8 7 9 0 4 3 3
5 5 6 8 4 7 9 4 3 8 9 2 5 7 5 4
9 3 1 9 2 1 7 0 4 4 5 6 1 1 8 5
1 8 1 2 7 7 8 3 8 6 4 2 5 4 5 6
8 2 7 4 5 7 0 8 7 6 1 1 0 5 3 2
9 4 0 4 1 1 6 4 6 5 5 5 5 2 9 1
0 4 6 8 4 2 7 6 5 3 3 3 6 6 0 0
4 3 7 1 3 4 5 4 3 5 4 9 1 5 8 4
```

- ☐ 836496
- ☐ 097859
- ☐ 858219
- ☐ 327610
- ☐ 283504
- ☐ 407129
- ☐ 072343
- ☐ 263857
- ☐ 099933
- ☐ 817448
- ☐ 573772
- ☐ 333567
- ☐ 448196
- ☐ 819504
- ☐ 378358
- ☐ 389257

Puzzle 39

```
4 7 9 4 7 6 5 5 2 6 1 3 6 9 6 7
3 4 5 4 7 3 7 5 3 1 8 3 9 1 2 3
4 2 4 7 5 5 9 3 8 1 6 8 8 1 2 2
5 4 2 5 2 1 9 3 5 9 8 0 0 5 1 8
6 5 1 1 7 6 4 1 5 9 2 2 5 9 6 7
0 6 8 7 1 2 3 0 4 8 7 5 1 1 0 8
3 6 8 3 6 2 8 3 1 6 6 2 9 9 7 8
7 7 0 6 7 1 5 9 2 0 0 9 1 0 1 0
7 4 7 3 9 3 1 8 1 1 4 2 5 0 8 6
0 4 5 4 4 0 2 9 2 1 5 6 3 4 7 2
2 2 0 7 8 0 8 8 7 6 2 4 8 7 2 2
7 6 7 2 5 0 5 8 1 8 7 8 0 4 9 6
7 5 8 1 2 4 7 9 0 8 9 6 6 1 8 5
5 6 3 2 9 6 7 6 4 1 9 5 5 7 0 1
6 8 3 0 3 5 4 0 8 7 5 9 6 8 0 5
8 1 5 7 3 9 6 1 5 6 9 2 7 6 4 2
0 4 1 3 7 7 7 1 2 0 5 0 8 0 6 8
7 8 1 5 0 8 5 5 7 5 7 2 7 7 6 2
```

- ☐ 052767
- ☐ 808876
- ☐ 806567
- ☐ 479476
- ☐ 025292
- ☐ 736347
- ☐ 215634
- ☐ 322112
- ☐ 451401
- ☐ 706122
- ☐ 566744
- ☐ 123362
- ☐ 974218
- ☐ 455832
- ☐ 813573
- ☐ 279959

Puzzle 40

```
6 8 8 3 8 0 9 6 8 1 4 5 0 4 5 1
0 5 0 2 5 2 4 2 8 6 8 1 8 7 5 8
6 4 4 8 4 1 8 2 1 7 0 7 0 6 8 1
9 7 6 5 9 2 4 2 1 4 3 5 2 2 0 0
9 4 8 5 1 3 6 9 5 3 9 3 4 0 1 3
5 1 0 3 8 8 3 1 1 7 7 8 4 1 8 7
6 6 3 4 8 0 0 6 4 2 4 1 2 7 8 2
9 4 0 6 2 5 6 7 7 9 5 4 6 4 2 8
5 5 9 2 1 0 8 4 3 9 4 3 9 6 4 1
4 2 3 8 2 1 1 4 9 9 7 9 9 2 7 2
9 1 5 9 5 1 3 3 8 5 1 6 1 9 3 8
3 2 2 9 5 2 3 0 8 5 6 7 4 0 2 9
4 7 8 2 5 5 5 4 2 6 8 6 6 1 3 8
2 8 0 3 8 7 7 3 4 5 9 8 4 8 3 6
8 7 4 1 6 9 2 9 3 4 2 5 9 0 0 4
3 5 6 2 3 6 6 2 6 2 6 5 8 7 2 6
2 7 6 2 0 4 6 3 4 8 0 8 2 0 7 0
3 6 7 5 5 4 6 5 4 1 1 9 9 6 0 5
```

- [ ] 361130
- [ ] 602213
- [ ] 843943
- [ ] 583876
- [ ] 341242
- [ ] 255542
- [ ] 084364
- [ ] 236626
- [ ] 952439
- [ ] 442080
- [ ] 513695
- [ ] 418217
- [ ] 142460
- [ ] 926471
- [ ] 869083
- [ ] 649852

Puzzle 41

```
9 8 1 7 6 0 6 5 7 4 2 0 2 5 0 9
0 8 1 3 1 5 6 1 9 3 4 6 3 6 2 4
4 0 3 8 0 7 5 8 4 7 4 6 1 5 9 8
1 4 4 2 0 9 4 6 0 9 6 9 1 5 8 8
3 9 1 6 0 8 9 9 2 0 0 8 6 9 9 9
9 6 7 4 9 4 4 2 4 2 3 3 6 0 2 7
3 0 2 9 0 4 6 6 5 8 6 1 9 7 3 0
8 7 6 1 6 7 6 1 4 1 1 3 7 4 7 1
3 0 4 5 5 7 1 6 8 5 8 6 7 9 9 8
6 2 4 0 0 4 2 0 7 4 6 7 4 4 1 1
3 0 6 4 5 7 9 9 7 8 4 6 0 4 8 2
8 8 2 9 6 2 8 3 4 3 8 4 3 0 7 2
2 7 1 1 6 9 1 5 2 2 7 6 1 3 2 6
9 1 8 7 4 9 9 0 2 2 2 6 7 4 7 4
6 2 5 1 7 9 3 4 0 5 1 2 6 7 8 3
6 2 0 5 7 9 9 7 3 8 6 9 9 3 1 5
0 2 4 2 1 4 3 2 0 9 5 8 9 3 2 9
0 4 0 2 6 6 2 0 1 5 0 6 1 1 7 0
```

- ☐ 420746
- ☐ 341242
- ☐ 491715
- ☐ 363839
- ☐ 200869
- ☐ 685867
- ☐ 167616
- ☐ 423360
- ☐ 152276
- ☐ 832046
- ☐ 784604
- ☐ 087122
- ☐ 474615
- ☐ 237918
- ☐ 916856
- ☐ 299031

Puzzle 42

```
1 5 9 6 1 1 3 5 2 2 4 1 7 1 4 7
8 8 0 0 1 9 9 8 7 1 4 1 8 1 5 7
4 8 8 5 5 8 5 2 6 2 5 8 0 4 7 6
9 7 6 5 9 4 3 7 3 0 7 9 0 3 0 2
4 3 0 9 8 4 1 2 1 9 4 2 8 9 9 7
9 7 9 1 3 9 2 4 5 9 2 8 1 7 7 1
8 2 2 0 7 5 1 5 2 8 5 3 8 4 3 6
7 4 1 7 6 2 0 0 9 3 5 1 2 2 0 3
3 3 8 3 7 0 8 9 1 7 9 0 5 4 3 1
4 5 2 1 4 3 5 9 2 8 8 9 1 2 3 8
8 0 1 1 3 3 5 6 2 8 2 6 9 2 1 3
1 2 0 5 6 2 2 7 5 8 1 3 2 5 2 4
5 9 3 3 0 2 5 6 7 7 9 1 5 2 2 3
4 6 9 5 5 4 5 6 5 9 3 9 0 4 6 9
7 4 4 2 2 9 0 0 2 6 8 4 3 9 9 8
3 2 3 1 6 1 8 0 6 5 5 0 6 2 5 9
4 3 8 5 2 2 5 1 7 3 1 4 1 7 0 3
2 2 5 7 5 0 9 5 7 9 6 1 6 2 3 6
```

- [ ] 127275
- [ ] 634767
- [ ] 495679
- [ ] 978622
- [ ] 326565
- [ ] 349301
- [ ] 764375
- [ ] 625855
- [ ] 033122
- [ ] 378951
- [ ] 931979
- [ ] 214890
- [ ] 328101
- [ ] 481547
- [ ] 515917
- [ ] 522517

Puzzle 43

```
1 3 4 6 1 2 8 0 2 7 4 3 0 0 9 4
8 2 0 5 5 8 7 4 1 0 3 2 7 6 7 9
7 3 9 7 7 4 9 7 3 0 8 1 9 7 0 2
1 5 4 8 7 8 5 3 7 2 7 9 2 7 9 6
0 5 8 5 5 0 2 4 4 4 1 8 0 6 5 3
6 1 8 0 7 9 9 8 2 7 3 4 3 3 8 6
1 2 8 8 9 5 1 4 3 0 2 8 2 2 4 6
3 9 6 1 6 2 0 7 8 3 4 4 8 9 0 4
3 8 4 2 1 1 9 4 0 3 4 4 5 7 1 2
5 2 1 3 6 7 0 9 0 0 4 9 7 7 9 8
0 4 1 8 2 9 9 1 4 0 2 6 9 9 1 9
8 2 4 1 7 5 4 3 8 1 2 1 8 3 4 0
5 1 4 4 1 2 3 5 3 8 0 0 6 4 6 6
9 5 5 3 8 6 5 6 5 4 8 3 5 4 7 7
1 6 9 4 7 9 5 9 7 5 7 0 7 8 6 8
7 9 0 9 3 7 6 6 0 7 0 7 6 1 3 4
1 5 5 9 3 2 6 6 0 2 8 6 3 2 4 2
2 3 7 4 4 2 1 8 2 7 5 1 4 4 5 5
```

- ☐ 081444
- ☐ 016885
- ☐ 908482
- ☐ 703300
- ☐ 834571
- ☐ 698444
- ☐ 783448
- ☐ 719580
- ☐ 301478
- ☐ 410371
- ☐ 677342
- ☐ 538006
- ☐ 992908
- ☐ 660707
- ☐ 711810
- ☐ 397792

Puzzle 44

```
4 4 1 1 6 8 5 6 2 3 0 4 6 8 3 6
1 1 9 6 9 6 2 2 6 6 3 3 4 7 3 5
6 8 9 8 3 4 7 8 9 6 3 4 5 7 0 8
8 3 6 9 1 5 2 0 5 2 1 6 4 8 4 7
0 4 7 1 3 3 4 7 2 5 1 6 4 3 3 4
3 4 4 7 6 3 0 5 4 0 1 5 7 6 9 7
6 1 8 4 1 8 5 0 7 7 9 7 5 3 5 2
5 4 3 4 6 0 7 6 0 4 5 9 4 5 0 7
7 9 9 5 3 2 0 7 6 0 8 6 4 6 9 5
8 2 1 3 9 5 3 4 9 9 6 4 3 5 9 1
3 1 7 6 5 5 9 6 6 7 6 1 0 4 4 4
7 1 0 7 8 1 3 6 0 4 5 0 6 3 4 9
3 4 3 4 9 6 0 6 9 7 6 4 9 0 1 8
0 5 9 1 8 6 3 1 9 7 7 3 4 1 9 5
4 3 4 3 5 3 5 5 8 7 2 5 2 3 8 9
7 0 9 6 8 8 1 3 5 4 5 5 3 9 7 3
1 0 6 8 5 5 3 3 8 6 6 0 9 8 0 4
8 3 0 8 0 2 3 0 0 6 2 0 9 9 5 1
```

☐ 985936 ☐ 003189 ☐ 579906 ☐ 386403

☐ 315303 ☐ 653399 ☐ 779754 ☐ 336622

☐ 354574 ☐ 583661 ☐ 574724 ☐ 093364

☐ 454475 ☐ 944198 ☐ 418344 ☐ 043950

Puzzle 45

```
8 1 1 9 3 7 6 4 1 8 4 6 3 3 8 2
1 1 2 1 1 4 2 0 1 8 4 5 1 9 9 1
6 6 1 8 0 4 9 3 5 1 1 7 6 6 2 1
7 0 8 2 5 7 4 0 3 9 2 8 7 0 8 6
7 5 9 1 1 7 6 6 8 3 0 2 3 3 2 5
7 2 9 5 5 1 6 0 8 4 7 8 7 3 2 0
3 5 4 6 2 3 6 9 6 4 4 8 1 3 6 8
5 2 2 0 9 2 7 2 7 8 0 4 1 0 7 5
7 9 1 1 1 9 6 2 9 0 0 0 6 1 2 5
0 7 6 5 2 4 2 4 2 5 5 8 6 0 3 7
8 1 4 7 3 5 6 9 3 4 0 2 6 8 5 1
2 7 9 8 4 6 7 7 3 7 8 5 0 3 0 6
1 1 6 7 6 3 7 3 4 5 5 5 4 0 8 8
2 6 8 8 2 8 9 6 9 4 8 4 4 3 4 0
2 0 6 5 6 8 2 6 7 3 0 7 4 0 6 9
7 5 3 4 6 3 3 4 6 8 9 4 9 0 8 7
5 7 0 9 7 2 9 5 1 2 0 8 3 4 2 6
1 6 0 8 9 3 7 7 8 4 7 2 1 4 6 2
```

- ☐ 984677
- ☐ 759117
- ☐ 673711
- ☐ 908444
- ☐ 097295
- ☐ 144684
- ☐ 830300
- ☐ 629000
- ☐ 085805
- ☐ 242425
- ☐ 065128
- ☐ 935117
- ☐ 243802
- ☐ 653741
- ☐ 206568
- ☐ 672350

Puzzle 46

```
5 4 3 9 0 0 1 3 0 7 7 2 2 6 2 7
1 4 0 3 6 6 6 6 3 4 5 3 0 1 1 8
6 8 1 6 7 7 5 6 3 4 8 8 1 7 8 3
6 7 5 1 8 1 6 8 9 5 7 1 8 9 9 0
3 4 5 7 5 5 6 9 1 1 8 6 3 5 5 7
8 8 0 8 8 2 2 4 4 9 4 8 0 7 1 5
2 2 1 4 3 3 8 4 7 4 2 4 9 8 8 8
0 5 0 4 0 9 1 2 8 1 5 9 8 4 4 6
0 2 6 5 6 9 9 6 3 9 1 0 1 0 3 8
9 7 7 3 5 9 2 4 9 4 9 3 2 6 4 1
1 5 2 8 0 6 3 0 2 4 5 6 3 6 5 1
0 9 7 4 3 8 0 7 8 5 0 2 3 7 2 2
8 7 7 0 5 2 6 5 1 7 9 8 2 5 1 7
8 2 9 4 1 5 6 2 5 6 2 5 8 1 5 8
7 4 8 4 5 7 0 6 1 8 1 7 8 1 1 1
5 8 4 1 8 9 8 6 2 0 2 2 9 8 3 4
0 6 5 2 1 9 6 1 8 2 7 4 0 8 1 6
6 5 6 9 1 2 8 3 8 7 3 8 9 7 7 6
```

- [ ] 965575
- [ ] 381469
- [ ] 675118
- [ ] 710759
- [ ] 543282
- [ ] 291826
- [ ] 889424
- [ ] 807850
- [ ] 100934
- [ ] 256258
- [ ] 226623
- [ ] 724550
- [ ] 675181
- [ ] 912560
- [ ] 819291
- [ ] 577618

Puzzle 47

```
4 5 9 3 2 1 2 1 6 0 4 0 8 2 0 4
5 3 0 3 7 2 6 6 4 4 3 8 9 6 4 7
1 0 6 1 8 7 9 6 7 6 2 7 4 3 0 6
9 4 3 6 3 8 6 2 2 3 3 1 0 8 6 2
9 7 1 1 4 9 4 0 0 6 4 1 8 9 9 9
0 2 7 8 4 9 2 6 6 7 5 8 0 1 2 5
2 7 6 8 7 2 3 2 0 6 4 0 8 8 3 0
6 8 1 0 5 7 2 1 0 9 0 0 4 9 5 7
0 1 0 7 0 8 3 9 9 2 4 9 4 9 4 2
6 7 6 9 9 2 6 4 3 4 8 3 8 5 9 0
2 0 6 9 2 3 9 6 4 3 0 9 2 0 7 2
4 5 6 3 9 0 4 3 1 9 9 5 4 1 9 3
9 4 4 6 8 5 9 6 4 9 1 7 3 7 1 9
9 2 5 7 0 8 1 0 5 8 1 8 8 7 4 8
1 8 9 2 7 7 3 7 2 3 4 0 5 3 7 1
3 6 7 3 8 8 0 0 5 2 0 5 9 3 2 7
6 6 0 6 3 7 1 6 6 3 7 9 9 5 1 2
0 8 1 6 3 1 5 3 7 0 3 6 6 7 7 9
```

☐ 982022 ☐ 081058 ☐ 194636 ☐ 844824

☐ 389189 ☐ 997088 ☐ 061879 ☐ 491853

☐ 409365 ☐ 711494 ☐ 454323 ☐ 238793

☐ 638622 ☐ 377606 ☐ 994260 ☐ 940537

Puzzle 48

```
7 2 9 5 2 4 7 9 1 4 6 7 3 2 6 7
4 2 5 7 2 6 1 2 5 8 0 8 5 0 7 6
7 0 1 3 5 6 4 7 9 4 1 9 0 4 9 1
5 0 3 0 3 3 1 4 6 8 9 9 1 3 8 9
2 3 6 4 5 3 4 8 4 9 2 7 4 1 0 3
7 0 6 7 0 4 7 2 0 2 6 3 2 9 5 2
8 0 3 5 1 5 9 2 7 1 8 5 3 3 6 9
0 5 9 6 3 7 8 4 1 6 4 2 4 9 0 2
8 6 4 0 1 9 1 9 4 8 3 8 0 3 2 0
8 4 7 2 2 7 8 7 3 6 0 2 2 0 2 5
6 0 8 8 1 7 4 6 0 5 2 1 0 3 5 5
7 5 2 8 7 9 1 5 9 4 4 3 7 3 6 7
8 8 4 2 7 0 0 9 9 8 5 8 0 7 5 3
4 7 8 6 3 8 7 4 4 7 3 2 7 1 8 0
1 3 2 0 0 7 7 5 4 1 1 1 8 4 4 3
6 0 3 8 4 3 2 8 8 0 2 7 0 8 4 1
3 6 3 0 1 0 2 6 1 2 7 4 9 5 1 4
4 0 1 6 9 5 5 8 1 5 9 7 1 3 4 2
```

- ☐ 088174
- ☐ 391340
- ☐ 784539
- ☐ 358172
- ☐ 565220
- ☐ 372180
- ☐ 786387
- ☐ 721620
- ☐ 350142
- ☐ 114577
- ☐ 484927
- ☐ 125808
- ☐ 067171
- ☐ 867841
- ☐ 135647
- ☐ 177304

Puzzle 49

```
4 2 5 9 8 5 7 7 4 7 0 3 0 9 1 9
7 0 5 7 4 3 4 2 0 3 4 1 4 0 3 1
4 4 5 9 9 9 3 9 9 9 3 7 8 9 1 5
9 5 9 6 3 2 3 6 7 1 4 5 4 8 3 5
9 9 0 5 7 1 2 9 6 4 0 5 3 5 8 5
8 8 1 8 9 3 9 6 3 1 5 4 5 8 1 2
0 0 9 6 9 2 6 0 4 6 2 5 2 4 6 7
0 4 1 9 1 7 1 4 7 0 4 7 2 9 4 5
2 8 9 6 3 1 0 8 1 1 5 3 0 1 2 5
8 8 3 4 2 2 8 2 9 3 8 8 7 6 2 0
1 2 3 8 2 3 2 4 9 2 0 1 8 7 3 3
6 3 3 1 7 3 0 1 6 9 6 2 6 9 1 5
0 5 0 9 3 7 5 1 5 1 0 0 7 4 2 5
2 0 5 6 0 3 7 5 2 6 3 2 2 4 7 2
0 7 5 3 6 0 6 9 4 0 8 1 5 0 4 0
9 7 0 8 6 5 1 2 6 9 1 2 3 9 4 8
0 7 3 6 6 5 4 4 5 0 9 4 8 6 0 6
2 0 3 1 4 6 1 4 1 8 2 7 8 7 0 4
```

- ☐ 369398
- ☐ 640629
- ☐ 399954
- ☐ 069408
- ☐ 823507
- ☐ 060854
- ☐ 037526
- ☐ 381202
- ☐ 631316
- ☐ 420341
- ☐ 207867
- ☐ 532494
- ☐ 044976
- ☐ 989932
- ☐ 296405
- ☐ 714548

Puzzle 50

```
4 5 7 7 9 6 9 2 3 3 5 9 7 0 7 8
7 6 5 6 0 1 7 8 6 4 0 0 7 3 7 6
1 3 4 7 6 0 0 7 1 1 1 7 7 7 5 0
8 1 4 5 5 1 9 5 3 1 4 8 6 8 0 7
3 8 2 5 3 7 9 5 8 6 6 1 7 9 9 6
1 8 2 9 9 7 3 3 3 8 1 3 6 7 1 5
0 8 2 2 6 9 5 3 1 2 4 1 6 1 6 7
8 5 0 3 3 4 0 7 9 1 1 1 0 9 1 8
3 5 9 7 4 8 7 7 9 0 6 5 0 2 5 7
9 9 7 3 9 5 0 6 7 8 2 0 2 1 4 2
5 0 5 9 1 3 1 4 8 8 6 3 3 6 8 3
3 8 5 8 6 8 5 3 7 1 5 9 8 0 4 7
8 9 9 3 2 9 4 4 3 5 9 0 5 6 1 2
3 4 6 8 7 7 7 8 6 8 4 5 9 7 5 2
8 8 3 1 4 2 4 5 9 2 2 5 7 7 4 0
0 6 6 7 2 5 4 3 9 4 2 7 8 4 3 8
1 6 0 7 6 7 0 5 0 5 0 3 6 8 1 9
5 6 3 5 4 3 3 9 8 6 0 2 4 1 5 3
```

☐ 757339	☐ 332969	☐ 235900	☐ 905877
☐ 361102	☐ 319505	☐ 574083	☐ 589845
☐ 636118	☐ 631888	☐ 879583	☐ 584346
☐ 577711	☐ 391667	☐ 516190	☐ 606774

Puzzle 51

```
6 4 3 7 3 6 4 5 9 3 6 0 4 5 6 0
0 8 5 2 5 6 3 0 9 8 1 5 7 8 1 0
2 4 5 4 6 4 5 5 7 1 0 3 8 0 1 5
2 2 4 6 7 2 2 7 2 5 0 8 6 0 0 8
0 2 8 6 1 8 0 9 2 7 0 9 5 1 9 4
8 6 6 9 3 1 4 6 3 8 1 0 6 8 4 5
5 7 2 4 7 8 3 3 5 0 0 6 7 3 5 8
2 1 5 7 0 5 4 2 8 4 3 9 9 0 6 3
7 5 9 8 6 5 9 5 1 7 4 5 3 5 9 8
5 0 0 5 3 3 3 5 6 0 2 0 0 2 7 3
7 3 0 1 3 7 9 5 6 4 8 5 0 8 8 8
6 3 7 9 5 8 2 7 3 2 2 0 8 4 7 1
9 5 1 3 3 5 0 6 6 7 4 6 8 2 0 2
6 9 4 0 2 9 2 9 3 7 7 1 9 1 9 8
6 8 7 6 5 3 9 9 7 8 4 7 6 6 8 7
9 6 4 5 1 8 3 4 2 1 4 8 7 3 7 0
6 7 6 5 4 2 0 9 4 9 5 0 5 4 0 6
1 0 8 3 8 0 7 3 0 4 2 0 7 1 6 1
```

- [ ] 624654
- [ ] 179083
- [ ] 818246
- [ ] 265957
- [ ] 823908
- [ ] 535377
- [ ] 737841
- [ ] 098157
- [ ] 303508
- [ ] 700939
- [ ] 527169
- [ ] 255105
- [ ] 762248
- [ ] 537323
- [ ] 800183
- [ ] 736459

Puzzle 52

```
9 1 0 4 2 3 6 4 9 7 9 4 7 1 7 9
6 0 5 7 2 5 7 9 7 6 8 1 9 2 0 8
6 3 1 2 9 8 7 0 0 8 0 6 6 7 0 7
2 1 5 0 5 6 8 9 0 1 5 5 8 2 3 7
0 6 2 9 2 7 8 4 7 1 4 0 7 3 6 4
1 2 1 5 7 2 0 9 7 5 6 0 7 5 8 5
8 4 4 3 5 7 5 1 1 0 7 8 3 3 4 2
3 2 2 0 7 7 6 1 9 1 6 8 5 8 6 7
7 0 6 0 4 3 7 1 0 9 6 9 6 9 6 1
9 1 0 9 2 7 2 2 9 3 6 5 3 9 6 4
3 6 8 4 4 6 5 2 6 4 0 9 0 0 8 0
1 3 5 0 7 9 5 2 3 9 4 7 0 9 9 9
7 7 6 6 8 1 3 9 6 6 6 3 1 0 2 3
1 9 6 2 0 9 0 6 8 7 6 2 5 2 6 3
7 3 9 7 4 6 0 8 9 4 2 5 2 4 2 1
5 0 0 7 0 6 7 3 5 2 8 2 8 9 2 4
9 0 1 7 7 4 2 2 3 4 9 9 8 9 9 6
6 6 0 2 1 1 1 6 9 8 0 9 5 6 7 0
```

- ☐ 252678
- ☐ 903960
- ☐ 672773
- ☐ 631298
- ☐ 900359
- ☐ 656067
- ☐ 749794
- ☐ 008066
- ☐ 116509
- ☐ 901774
- ☐ 714073
- ☐ 317175
- ☐ 018379
- ☐ 636909
- ☐ 966631
- ☐ 725922

Puzzle 53

```
0 3 4 4 2 4 8 0 2 8 9 0 3 1 9 8
4 7 6 6 3 6 5 2 9 4 5 5 5 0 2 4
7 1 3 9 8 5 2 8 9 5 5 8 3 3 2 1
2 1 2 0 6 1 1 6 1 4 8 6 7 2 6 4
1 1 1 1 0 6 1 0 8 8 2 8 1 5 7 4
8 7 6 6 5 2 3 0 4 5 5 0 8 8 4 9
6 9 7 3 3 9 1 0 3 4 8 0 6 4 1 5
9 2 8 6 9 9 6 1 5 5 2 4 9 5 1 2
7 6 8 3 0 4 8 8 3 6 5 5 5 5 0 3
9 6 8 6 6 9 8 2 8 0 6 2 7 3 9 6
8 6 1 6 0 4 7 1 1 6 8 8 9 9 7 5
4 3 3 7 7 4 9 9 6 3 3 6 1 7 3 1
1 1 9 8 6 0 3 6 8 0 3 2 0 3 4 4
3 7 8 2 3 4 9 1 7 2 6 5 9 5 7 4
9 0 8 4 8 3 0 8 4 8 6 5 3 4 4 6
4 8 1 6 2 6 6 0 7 5 9 0 2 1 6 9
1 5 6 3 8 2 7 7 4 8 2 3 5 9 4 9
7 6 5 2 0 8 9 9 1 9 4 1 6 7 1 3
```

- ☐ 845485
- ☐ 726595
- ☐ 476636
- ☐ 219618
- ☐ 106108
- ☐ 011476
- ☐ 998861
- ☐ 356848
- ☐ 032584
- ☐ 665230
- ☐ 866982
- ☐ 193379
- ☐ 337749
- ☐ 748235
- ☐ 048836
- ☐ 558258

Puzzle 54

```
6 9 2 9 9 1 7 9 8 3 0 5 0 3 1 7
3 9 2 8 2 3 3 7 6 3 1 0 2 5 3 7
5 7 5 0 7 2 6 0 5 1 4 9 4 8 6 7
7 9 0 6 9 4 0 2 4 0 6 5 2 4 9 1
2 3 9 7 5 9 4 4 7 1 1 8 2 0 5 4
7 4 4 3 2 2 2 3 4 9 3 4 8 5 2 1 0
4 2 4 6 5 6 4 0 5 3 4 0 6 6 9 9
8 0 3 4 7 4 0 9 4 9 7 5 9 2 4 5
2 1 2 9 6 2 4 5 7 7 9 0 8 6 1 9
9 8 1 2 5 0 8 5 6 1 6 9 2 1 7 5
4 1 2 1 8 2 1 2 4 7 0 9 2 6 3 5
6 8 1 5 8 4 8 0 6 6 9 8 4 3 3 9
5 6 8 6 7 1 1 0 3 6 3 0 5 8 3 9
0 9 3 1 0 5 7 5 1 3 7 4 6 6 0 9
2 6 3 5 0 4 4 0 1 8 6 6 8 7 9 2
3 5 0 4 5 6 3 1 3 0 8 6 1 6 1 4
8 6 1 4 3 4 9 1 7 6 1 3 5 5 5 5
8 7 5 9 3 9 4 2 8 7 0 7 2 5 6 9
```

☐ 575072   ☐ 798305   ☐ 865422   ☐ 631025

☐ 904743   ☐ 225094   ☐ 658052   ☐ 624577

☐ 616262   ☐ 928472   ☐ 849415   ☐ 960188

☐ 273458   ☐ 616803   ☐ 322349   ☐ 951941

Puzzle 55

```
6 7 2 9 5 4 2 3 2 5 5 3 6 5 8 7
7 4 6 5 9 8 8 4 2 4 0 8 0 1 9 8
4 5 3 7 0 5 4 4 0 2 0 7 4 9 0 9
3 8 1 8 7 5 2 5 5 6 5 0 6 6 1 6
8 7 2 2 0 2 9 2 9 7 6 4 5 7 9 6
4 4 2 0 3 1 7 2 3 0 0 7 0 8 8 9
9 9 5 5 5 4 1 7 8 2 6 8 3 5 2 5
8 1 7 8 9 4 8 5 3 0 6 3 6 1 4 5
5 0 7 3 1 3 3 7 3 5 8 3 9 0 5 1
2 6 8 6 1 3 5 5 3 0 2 7 6 2 6 6
7 4 2 1 3 8 4 4 6 3 4 9 7 0 6 4
5 3 4 9 4 1 6 5 7 1 7 4 0 9 0 2
0 6 1 3 6 3 6 4 4 3 9 0 3 9 8 9
1 4 2 4 7 1 9 2 5 7 6 7 0 0 7 6
0 1 8 6 1 5 8 8 3 8 9 0 3 3 4 3
7 5 6 8 9 4 8 7 2 6 9 6 9 8 0 2
9 0 6 9 5 2 0 0 5 1 1 7 5 2 0 7
8 9 5 3 1 2 6 5 4 3 2 4 5 9 3 2
```

☐ 636035 ☐ 529174 ☐ 507313 ☐ 348733

☐ 634970 ☐ 095485 ☐ 296265 ☐ 952232

☐ 428910 ☐ 726969 ☐ 005117 ☐ 631225

☐ 967851 ☐ 804248 ☐ 439039 ☐ 571740

Puzzle 56

```
9 9 0 4 2 9 9 6 1 5 0 0 3 2 3 1
9 5 5 7 1 8 3 3 1 6 7 1 7 7 9 2
6 9 4 0 0 3 6 1 0 5 9 5 8 1 7 0
8 8 5 4 0 0 6 3 4 5 9 3 4 7 2 0
8 0 4 3 0 9 6 0 1 6 3 4 8 2 0 5
1 3 1 0 3 7 9 1 9 5 8 8 2 2 2 9
1 2 3 5 4 2 3 1 5 9 7 3 7 6 7 5
1 6 4 0 5 3 0 2 5 8 9 6 7 3 8 7
3 5 6 2 4 4 5 6 0 9 7 5 7 5 8
8 7 3 1 3 9 6 1 6 1 7 2 0 7 8 3
1 0 7 5 5 3 7 0 4 4 0 8 3 0 5 7
8 8 7 8 1 2 7 2 2 1 4 3 8 8 8 3
0 0 1 9 3 5 8 9 5 3 5 6 0 9 0 1
3 2 6 2 7 3 7 3 0 6 9 2 7 7 7 9
3 6 7 9 4 6 8 8 2 5 3 1 0 6 8 8
4 9 4 9 6 4 3 2 5 3 4 6 5 7 1 2
8 9 8 9 8 1 9 6 0 2 5 7 0 3 9 5
3 2 5 3 4 4 7 8 4 0 4 4 1 4 8 4
```

- [ ] 276384
- [ ] 544073
- [ ] 852035
- [ ] 688111
- [ ] 562444
- [ ] 044083
- [ ] 469494
- [ ] 946882
- [ ] 272187
- [ ] 141365
- [ ] 784827
- [ ] 966639
- [ ] 798077
- [ ] 602570
- [ ] 432534
- [ ] 693137

Puzzle 57

```
8 3 4 3 8 9 1 4 6 7 3 2 0 8 2 1
2 6 6 5 3 0 0 2 2 0 3 3 9 1 4 7
1 8 8 6 0 6 1 5 9 5 8 6 0 9 2 7
1 0 7 9 6 3 6 2 9 9 0 5 5 4 5 9
2 0 0 4 1 6 6 7 2 9 5 8 6 3 6 7
0 2 1 8 0 7 7 9 6 6 7 4 7 5 1 1
6 5 2 3 7 0 4 4 5 3 4 7 6 5 5 2
6 9 4 9 4 6 4 8 9 7 1 3 1 4 3 4
9 1 7 3 1 6 5 0 2 8 4 9 9 5 5 8
3 8 9 7 7 9 4 1 4 8 8 9 2 3 6 5
7 7 7 2 5 2 7 5 0 8 2 2 1 5 0 3
6 3 8 4 6 0 3 4 0 3 5 4 1 4 4 0
7 3 5 1 7 7 8 7 9 3 4 6 9 1 2 4
4 4 8 4 6 5 1 4 2 0 1 4 7 8 6 2
1 6 6 7 7 8 8 7 2 1 2 4 4 1 2 2
4 2 6 7 9 4 6 3 0 9 7 0 8 3 5 6
9 4 0 8 8 0 8 5 5 8 7 3 5 4 8 2
5 3 4 6 0 1 5 1 9 6 7 8 8 6 4 9
```

- [ ] 100876
- [ ] 636299
- [ ] 433781
- [ ] 527508
- [ ] 318145
- [ ] 250875
- [ ] 004782
- [ ] 427393
- [ ] 763192
- [ ] 071762
- [ ] 464599
- [ ] 855873
- [ ] 749882
- [ ] 604262
- [ ] 146732
- [ ] 369574

Puzzle 58

```
3 8 2 5 8 4 6 8 5 9 5 8 0 4 7 4
3 1 3 4 4 2 0 7 6 2 7 1 3 8 4 8
8 3 3 8 9 5 4 9 6 1 1 4 2 7 2 3
3 1 4 3 0 7 0 5 3 3 9 7 2 0 9 6
5 9 0 7 6 7 3 6 1 6 3 0 9 1 5 2
6 4 7 1 7 1 6 6 0 1 1 1 4 6 8 8
0 9 3 4 4 5 2 7 6 9 0 7 3 9 4 3
5 3 0 0 9 2 1 6 6 4 1 6 0 3 2 2
9 4 7 0 1 8 8 6 1 8 3 3 5 5 2 8
9 2 7 6 0 8 0 4 4 1 7 2 9 4 7 1
5 4 6 5 2 0 8 9 9 3 0 8 8 0 0 9
0 5 7 6 9 3 2 7 9 6 9 1 1 5 1 0
6 5 0 7 6 5 1 0 9 7 2 1 4 2 8 7
8 9 5 9 8 8 4 0 3 0 5 3 6 9 7 4
3 4 6 1 5 3 1 4 7 2 0 3 8 1 6 1
5 2 7 1 0 7 8 3 3 8 4 6 2 4 1 0
0 6 7 9 1 7 1 6 4 6 2 3 1 0 0 8
7 1 1 6 0 3 2 0 4 9 9 6 5 4 3 9
```

- ☐ 118236
- ☐ 949131
- ☐ 416601
- ☐ 472038
- ☐ 080672
- ☐ 079466
- ☐ 694598
- ☐ 995065
- ☐ 507651
- ☐ 076318
- ☐ 493709
- ☐ 925045
- ☐ 326461
- ☐ 793350
- ☐ 701725
- ☐ 771528

Puzzle 59

```
5 6 2 4 4 1 8 0 8 2 1 9 5 0 3 5
9 1 5 7 8 0 5 1 8 4 0 4 2 9 7 0
1 4 6 9 1 7 0 3 7 7 1 3 4 1 3 6
2 3 1 9 9 6 0 3 2 1 7 2 3 5 9 9
9 8 6 6 1 9 3 0 0 8 8 8 8 9 5 4
6 8 8 6 6 8 5 5 1 7 6 6 5 7 2 6
0 0 7 7 8 7 7 9 1 4 5 6 0 3 5 3
9 0 0 2 5 5 5 5 8 1 6 5 5 2 9 8
1 5 7 6 7 3 3 9 6 0 7 7 3 0 9 5
7 1 1 1 8 7 7 4 4 2 4 0 1 6 0 7
0 9 4 7 8 2 9 0 1 8 3 8 0 3 9 5
4 4 1 7 3 2 3 0 1 8 8 1 1 8 6 2
5 8 4 6 5 5 1 6 3 5 4 3 5 6 7 0
0 8 5 7 8 9 9 2 5 4 4 1 5 1 4 5
9 8 1 1 1 5 3 5 9 9 0 0 8 0 3 0
6 4 0 7 0 2 0 1 8 3 6 1 4 0 1 1
1 3 0 6 4 6 6 4 5 1 3 0 7 7 5 6
2 2 8 8 0 7 4 0 0 7 3 9 6 1 0 2
```

☐ 787700 ☐ 410288 ☐ 619184 ☐ 756682

☐ 621744 ☐ 518404 ☐ 043097 ☐ 379519

☐ 481435 ☐ 948835 ☐ 185953 ☐ 351118

☐ 058342 ☐ 037713 ☐ 134769 ☐ 094782

Puzzle 60

```
7 5 5 4 9 8 9 6 7 2 4 4 0 9 4 5
2 0 6 2 4 8 4 4 2 4 7 5 0 5 3 4
1 8 7 0 9 8 3 4 4 3 0 9 7 5 1 6
7 0 7 7 7 4 6 9 3 6 4 8 7 7 0 7
4 8 8 6 0 5 1 3 0 8 4 5 3 9 8 7
2 0 8 6 1 7 6 8 9 4 7 4 6 8 4 6
8 9 2 2 5 5 4 2 3 5 3 8 4 0 3 4
5 6 6 2 6 6 5 8 2 0 1 1 9 6 4 5
0 6 1 8 2 1 7 4 0 4 0 6 7 8 1 9
7 4 9 9 5 0 5 6 0 6 3 4 1 4 0 9
0 4 9 9 4 4 4 5 1 9 2 4 5 0 8 1
8 3 3 7 7 0 2 2 0 2 1 0 6 6 5 0
2 5 1 7 9 4 2 5 2 3 9 0 2 3 7 7
5 4 9 7 9 3 5 9 2 1 7 1 0 8 6 4
0 0 7 7 2 3 6 5 3 6 6 4 9 0 1 7
4 9 8 7 6 4 0 8 3 4 1 5 7 7 6 1
1 5 2 3 9 6 0 8 6 2 4 5 3 9 2 2
9 5 5 5 3 6 2 2 1 1 6 7 1 7 1 4
```

- ☐ 466908
- ☐ 220773
- ☐ 814925
- ☐ 427130
- ☐ 014348
- ☐ 002390
- ☐ 612919
- ☐ 501615
- ☐ 470447
- ☐ 156254
- ☐ 217428
- ☐ 985481
- ☐ 077074
- ☐ 341577
- ☐ 806932
- ☐ 794637

Puzzle 61

```
9 0 9 6 9 1 9 6 6 1 2 8 2 2 3 1
8 5 2 6 5 7 1 5 7 9 3 3 2 7 8 5
6 6 8 7 7 4 9 2 7 0 1 5 7 5 1 3
7 6 5 0 8 9 1 5 4 7 3 3 5 0 7 3
0 5 1 1 2 5 2 0 0 7 8 5 9 7 4 3
9 7 2 2 7 8 1 6 1 1 1 4 3 6 5 5
3 5 2 9 2 3 6 1 9 9 3 0 1 4 0 2
2 7 3 7 1 6 7 2 7 2 3 8 7 1 1
0 4 5 9 4 8 1 2 3 0 2 7 5 3 8 3
7 7 3 0 7 6 0 3 8 6 1 7 9 1 8 9
3 5 5 8 9 0 8 5 9 9 1 7 3 1 3 9
3 8 3 1 2 5 6 5 2 5 0 1 0 5 9 5
1 8 0 6 6 7 2 9 2 7 0 1 4 6 5 6
5 2 6 6 5 7 1 7 7 9 9 9 7 6 2 9
0 5 7 3 0 7 4 2 6 2 5 8 6 6 2 2
1 0 4 0 8 7 2 3 9 1 5 8 9 5 9 2
0 4 2 5 7 1 5 7 8 1 1 0 5 1 6 1
1 6 6 9 6 5 6 2 5 2 8 1 2 2 4 0
```

- ☐ 989725
- ☐ 297908
- ☐ 353540
- ☐ 926508
- ☐ 071158
- ☐ 501187
- ☐ 656905
- ☐ 368605
- ☐ 023907
- ☐ 225938
- ☐ 779101
- ☐ 574758
- ☐ 216691
- ☐ 593047
- ☐ 792692
- ☐ 175240

Puzzle 62

```
4 2 5 9 8 8 5 8 0 3 4 7 8 1 5 8
3 5 6 6 4 1 4 3 8 1 6 7 9 0 8 2
7 3 2 8 2 8 1 7 8 1 1 8 2 5 6 6
0 4 0 5 6 0 9 7 1 5 3 0 6 9 2 3
7 1 4 4 1 6 4 2 4 1 6 9 1 4 2 1
7 8 9 4 2 8 2 2 6 9 0 3 6 3 9 7
0 7 3 6 0 1 5 5 5 0 9 5 2 2 6 2
8 4 2 4 3 5 8 2 3 8 8 8 7 8 2 8
5 3 6 7 1 7 3 9 0 6 6 4 4 1 8 7
5 5 7 3 4 1 1 3 3 4 1 7 8 1 0 5
5 2 1 8 6 1 7 8 9 1 6 1 3 3 4 3
2 7 2 4 4 8 7 2 1 9 0 7 9 5 2 1
2 0 1 1 1 5 3 3 4 4 6 6 1 8 2 8
7 0 3 8 4 3 6 8 2 2 5 2 1 7 5 4
9 6 3 1 1 8 1 0 3 9 9 3 9 2 8 6
2 3 4 2 2 9 9 1 9 7 0 2 2 8 6 2
4 5 0 5 4 1 0 0 1 1 9 7 3 6 8 7
4 2 3 2 5 3 1 1 1 5 0 5 0 8 3 5
```

- ☐ 750117
- ☐ 489947
- ☐ 328534
- ☐ 425867
- ☐ 131415
- ☐ 095606
- ☐ 311823
- ☐ 261627
- ☐ 360155
- ☐ 430858
- ☐ 912784
- ☐ 221489
- ☐ 555227
- ☐ 807707
- ☐ 848062
- ☐ 183414

Puzzle 63

```
5 5 7 7 9 8 9 9 4 1 8 0 9 3 4 3
0 3 9 0 3 1 3 3 6 3 3 0 8 1 6 1
9 9 1 0 8 2 7 3 5 4 8 8 9 9 2 4
5 6 8 8 2 0 3 3 6 1 9 3 8 6 4 7
9 9 2 6 7 8 7 5 2 5 9 1 1 5 3 5
6 7 0 1 8 4 1 2 9 7 5 1 2 6 2 5
8 1 5 4 4 0 0 8 6 6 6 5 2 5 3 5
1 5 0 4 4 5 0 6 5 8 7 3 2 1 0 0
3 1 3 0 0 8 0 3 8 0 6 4 6 9 7 9
7 1 7 7 0 2 0 0 4 1 3 5 4 5 9 4
4 5 3 6 6 8 1 7 6 0 3 8 7 2 5 8
2 1 0 1 9 0 0 9 2 4 9 4 7 3 3 2
9 1 8 5 0 3 4 5 4 6 5 6 0 4 9 4
3 4 1 0 3 6 6 9 6 2 9 8 3 7 5 5
9 7 4 7 2 1 7 4 3 5 7 7 8 4 9 2
0 9 5 7 9 9 3 2 6 3 2 3 6 1 7 0
0 7 5 8 3 1 7 0 3 4 9 5 2 1 8 8
0 3 9 6 8 4 9 1 7 6 5 6 9 9 0 2
```

☐ 597036   ☐ 693646   ☐ 367659   ☐ 430713

☐ 807268   ☐ 120457   ☐ 527830   ☐ 579932

☐ 115114   ☐ 208405   ☐ 473186   ☐ 926564

☐ 751431   ☐ 100001   ☐ 962983   ☐ 622218

Puzzle 64

```
2 7 8 3 7 6 1 0 9 3 0 9 0 6 0 7
9 7 7 4 8 5 3 5 9 8 1 2 3 9 3 6
3 5 8 0 6 1 8 9 2 6 2 1 2 1 4 5
4 6 1 7 2 0 8 3 3 2 6 1 3 5 1 4
1 3 0 1 1 4 8 6 6 4 0 7 6 1 6 4
9 5 0 9 5 0 2 1 6 1 4 4 7 5 8 1
2 2 8 0 7 0 3 3 2 2 1 7 2 6 6 6
8 2 1 3 9 2 7 2 9 4 8 7 6 3 2 7
3 9 6 9 1 2 5 6 9 2 0 6 9 1 9 4
5 7 4 3 9 9 6 7 2 1 1 4 0 0 7 0
4 9 6 4 4 9 8 1 3 9 3 7 7 4 8 2
1 4 8 1 4 4 7 0 4 1 3 7 8 1 2 7
2 8 1 5 2 0 6 4 1 4 2 6 3 7 0 2
5 3 5 0 6 8 6 4 9 9 0 3 7 6 4 0
5 1 1 9 0 3 1 5 8 3 2 6 8 9 4 9
4 9 7 7 1 9 8 1 8 3 1 6 9 2 5 2
9 5 4 3 5 6 7 3 5 0 9 8 3 8 0 7
0 5 1 0 7 0 1 2 0 1 3 8 8 7 9 6
```

- ☐ 971163
- ☐ 191493
- ☐ 060903
- ☐ 341509
- ☐ 748535
- ☐ 361326
- ☐ 837610
- ☐ 455214
- ☐ 832137
- ☐ 900144
- ☐ 804992
- ☐ 705115
- ☐ 833261
- ☐ 053765
- ☐ 310417
- ☐ 710329

Puzzle 65

```
8 1 3 9 6 9 2 5 2 8 0 0 0 8 1 7
3 6 9 6 5 3 0 4 8 8 0 5 6 6 4 7
9 2 1 5 6 6 8 4 5 5 7 4 5 8 4 9
6 3 5 2 6 6 9 3 3 8 6 2 5 6 2 5
6 4 0 8 1 2 3 2 7 6 0 4 9 7 4 5
2 3 6 1 1 1 1 2 6 2 8 7 0 9 7 8
4 6 9 0 8 9 1 1 0 7 2 2 5 6 7 2
0 9 6 9 1 8 4 5 7 6 5 5 2 9 6 9
4 8 9 6 1 9 3 1 4 6 5 3 3 6 4 0
8 7 4 6 7 6 2 3 8 2 6 2 4 5 4 3
3 0 3 4 7 9 1 8 1 8 5 2 5 5 3 0
1 9 0 9 1 3 9 6 2 0 1 3 2 1 4 9
1 6 7 5 3 2 8 2 9 9 3 9 3 6 0 7
7 0 5 2 9 7 3 1 8 8 6 7 7 3 8 8
8 3 9 7 9 0 7 7 1 2 0 0 7 0 0 8
1 0 4 3 2 6 2 7 2 2 0 3 9 0 5 9
5 8 6 7 3 5 2 5 3 7 3 1 5 6 0 7
8 7 6 5 8 7 4 2 8 2 3 9 1 5 3 2
```

- ☐ 926753
- ☐ 338625
- ☐ 272623
- ☐ 211154
- ☐ 673811
- ☐ 735253
- ☐ 666799
- ☐ 960519
- ☐ 048805
- ☐ 569697
- ☐ 557458
- ☐ 181166
- ☐ 307594
- ☐ 247764
- ☐ 692528
- ☐ 223970

Puzzle 66

```
4 9 8 9 0 0 7 0 6 3 9 0 5 8 6 6
8 2 0 2 1 2 2 6 8 1 8 7 8 5 2 5
0 6 4 0 6 5 8 6 7 9 4 3 1 8 9 0
6 1 7 0 3 0 7 0 5 7 9 9 3 0 0 0
0 3 5 0 3 0 6 0 2 1 2 1 5 2 4 2
9 9 9 3 0 9 1 2 8 3 2 2 1 2 7 4
6 7 7 8 7 9 2 1 2 7 6 8 7 5 5 3
1 7 3 0 1 2 1 4 7 3 6 3 5 9 0 3
4 3 0 4 4 3 4 7 8 8 9 8 7 4 8 6
9 9 2 9 5 9 9 7 9 1 3 6 8 7 4 2
3 7 6 3 5 9 3 0 8 5 0 8 7 7 4 7
7 0 9 7 1 1 9 2 8 1 5 5 8 1 5 2
8 9 6 5 9 4 0 0 2 9 7 0 8 9 3 2
3 7 1 5 5 0 7 0 6 7 3 6 3 8 1 9
5 3 9 0 0 2 9 3 2 8 1 0 0 6 5 0
7 9 6 3 2 1 1 7 6 3 0 7 6 3 3 5
4 6 5 1 5 9 7 2 8 9 6 7 1 8 8 0
8 4 4 3 5 9 5 0 8 6 1 3 1 5 4 7
```

- ☐ 215620
- ☐ 795503
- ☐ 280236
- ☐ 153724
- ☐ 622396
- ☐ 914021
- ☐ 830671
- ☐ 301178
- ☐ 394079
- ☐ 878757
- ☐ 508445
- ☐ 289671
- ☐ 070098
- ☐ 696196
- ☐ 709739
- ☐ 818785

Puzzle 67

```
5 9 8 5 7 3 2 7 8 2 4 7 5 4 9 1
7 1 2 5 8 9 6 3 6 1 6 7 5 3 5 2
1 4 6 3 2 2 5 3 3 4 5 4 3 2 6 1
5 2 2 3 2 4 3 8 7 5 1 9 8 9 1 1
4 8 8 3 6 6 6 5 4 3 6 3 5 4 8 8
7 5 4 0 4 2 8 3 8 0 4 3 7 8 5 9
1 9 5 7 6 7 7 5 5 7 5 3 5 2 0 0
4 3 6 7 1 9 2 8 1 5 5 2 4 5 0 0
4 5 6 7 8 3 8 7 7 6 9 4 8 0 4 2
0 2 8 1 3 4 8 0 2 2 9 3 0 5 0 1
2 5 5 8 8 8 6 7 7 7 1 4 6 6 5 2
0 8 6 0 4 8 4 8 5 8 4 2 5 9 5 9
3 2 7 6 2 0 9 6 6 9 2 8 9 0 0 2
4 2 7 4 0 1 5 6 9 2 0 6 9 0 3 6
9 2 7 1 1 5 4 2 9 9 3 1 1 6 8 1
2 5 6 8 0 4 3 3 2 6 5 0 1 5 9 7
9 6 0 5 8 3 5 6 2 0 4 7 8 7 3 9
1 6 8 7 0 8 0 0 0 3 8 7 5 7 5 1
```

- [ ] 891578
- [ ] 793488
- [ ] 155245
- [ ] 025357
- [ ] 232685
- [ ] 727569
- [ ] 302044
- [ ] 574287
- [ ] 253958
- [ ] 963616
- [ ] 276209
- [ ] 274015
- [ ] 154299
- [ ] 659911
- [ ] 146322
- [ ] 727432

Puzzle 68

```
6 7 5 3 5 1 4 3 2 4 0 6 4 9 6 5
6 3 3 9 0 6 4 1 9 7 9 1 3 6 0 1
3 7 1 3 4 6 5 2 5 9 9 0 9 7 9 8
2 0 7 4 5 3 6 1 4 8 6 8 8 7 3 5
6 8 7 9 4 3 1 8 8 8 0 2 7 9 1 9
3 5 8 1 5 8 4 4 8 5 1 8 4 4 3 2
1 8 0 4 6 1 6 1 9 5 9 6 4 5 4 0
7 3 1 5 0 0 7 1 5 5 7 1 7 7 2 4
4 8 3 7 6 1 7 5 9 1 0 2 6 0 1 6
1 4 1 3 6 0 8 7 7 3 9 7 7 8 1 6
1 3 2 1 7 0 9 1 6 4 5 7 2 6 7 7
2 8 0 7 9 8 9 2 3 4 6 5 7 5 2 2
9 3 3 9 0 3 7 2 5 9 4 6 2 7 1 1
5 1 0 5 7 9 7 6 0 8 1 9 8 2 9 3
1 4 5 9 5 7 7 6 0 4 5 8 3 9 9 6
6 8 8 4 3 8 4 2 4 4 3 8 8 0 8 6
1 7 4 5 0 2 3 8 1 2 4 0 1 5 3 0
3 6 9 6 8 7 8 0 1 3 1 4 1 9 1 8
```

☐ 088965	☐ 158084	☐ 495273	☐ 093134
☐ 850302	☐ 545606	☐ 795988	☐ 419439
☐ 136087	☐ 858073	☐ 218411	☐ 542699
☐ 579760	☐ 406775	☐ 643298	☐ 663381

Puzzle 69

```
5 5 3 7 5 8 2 0 3 1 2 1 2 8 2 7
1 4 6 2 2 0 4 2 6 9 9 0 5 8 1 4
2 4 3 8 1 1 8 2 3 7 5 4 7 3 2 1
8 2 8 0 5 9 6 6 5 2 0 8 5 3 4 5
6 1 9 7 0 0 3 2 1 2 2 6 5 2 3 1
0 2 4 8 2 9 8 8 2 4 8 4 2 0 5 5
9 2 9 4 4 8 2 5 8 8 4 0 7 5 8 3
9 1 1 1 6 4 1 7 3 5 4 3 2 8 5 6
6 2 9 7 1 0 8 1 2 8 3 4 8 2 4 4
1 5 7 3 4 1 6 6 1 1 3 3 5 7 8 6
0 3 0 2 8 3 0 1 1 2 1 5 5 1 3 0
4 8 5 9 6 9 5 5 4 8 4 9 3 4 9 6
2 6 8 2 3 9 8 2 8 6 3 1 1 2 8 8
5 6 2 3 3 2 6 6 1 8 9 0 0 6 0 1
6 2 2 5 2 1 9 2 3 2 8 6 9 9 6 0
1 6 0 9 6 3 8 9 4 5 8 7 9 8 0 4
4 9 3 5 3 2 4 1 2 6 4 7 5 7 5 5
4 9 0 8 6 9 5 3 9 5 7 9 3 5 4 1
```

- [ ] 239828
- [ ] 614696
- [ ] 870827
- [ ] 534212
- [ ] 970582
- [ ] 058593
- [ ] 582714
- [ ] 174352
- [ ] 211182
- [ ] 508614
- [ ] 111926
- [ ] 638945
- [ ] 344820
- [ ] 383532
- [ ] 816844
- [ ] 221244

Puzzle 70

```
9 5 0 1 3 0 2 5 4 4 7 4 6 4 1 7
3 3 7 2 0 5 7 6 5 6 1 5 9 1 5 5
3 6 7 8 8 8 2 9 2 8 7 6 7 3 2 5
4 1 5 3 1 0 6 1 2 8 5 5 0 0 6 8
2 2 3 8 3 0 7 9 0 5 6 4 7 1 6 5
6 6 7 4 0 8 6 6 2 8 2 0 8 0 8 7
3 3 3 9 9 2 6 1 0 3 0 3 8 9 5 9
7 5 7 4 4 7 7 0 3 3 1 1 4 7 2 8
9 1 1 8 3 8 1 1 5 2 3 4 0 5 3 1
0 9 7 7 2 5 6 9 2 7 9 6 4 9 2 0
3 0 0 6 4 6 4 7 8 6 1 4 2 9 8 8
3 9 9 6 5 3 3 3 3 0 3 8 9 9 7 7
3 7 7 7 1 8 6 0 1 1 6 4 1 6 5 6
7 7 0 2 3 3 4 1 9 1 1 3 9 4 5 6
3 4 3 3 8 4 1 9 4 6 8 9 1 2 7 9
0 1 2 8 0 0 5 6 7 0 4 0 6 3 9 2
7 2 3 6 4 7 8 1 3 7 8 2 1 6 8 0
3 7 3 1 2 7 6 0 8 7 9 7 8 5 4 8
```

☐ 461766	☐ 327601	☐ 429191	☐ 975999
☐ 765615	☐ 361316	☐ 678494	☐ 485671
☐ 841946	☐ 744757	☐ 735770	☐ 668523
☐ 103038	☐ 348464	☐ 147790	☐ 362163

Puzzle 71

```
3 9 5 4 8 1 0 3 7 0 4 4 5 2 6 3
7 6 8 5 0 4 0 0 7 9 1 4 0 2 7 0
0 5 6 8 0 3 5 6 3 1 6 4 2 1 7 2
5 5 3 5 9 6 5 3 4 5 8 1 9 9 5 8
6 9 0 8 5 8 1 8 4 3 5 1 8 7 0 4
0 9 9 7 0 2 2 4 6 2 1 6 5 3 5 4
4 3 4 3 4 0 0 4 6 3 2 9 7 5 2 1
2 3 7 3 0 8 7 5 5 3 5 2 3 7 6 8
0 9 4 0 6 3 3 2 6 5 6 3 7 1 6 4
1 5 2 7 7 7 3 7 5 7 4 1 4 5 7
9 3 3 7 5 7 1 3 3 8 0 5 2 2 4 7
5 6 7 3 1 3 2 7 9 8 3 0 3 3 0 5
6 8 9 2 0 5 6 9 9 1 4 6 6 6 0 6
1 1 4 2 5 5 7 0 9 5 9 2 7 2 0 1
4 6 7 3 7 5 5 3 2 4 6 3 5 0 5 2
3 9 8 4 1 0 8 1 4 5 2 7 8 2 8 8
4 1 3 5 4 6 9 4 0 3 9 8 1 7 4 4
7 9 2 4 5 5 0 2 0 1 3 5 0 8 7 5
```

- [ ] 504007
- [ ] 316421
- [ ] 074833
- [ ] 599185
- [ ] 650062
- [ ] 777251
- [ ] 533055
- [ ] 370560
- [ ] 923640
- [ ] 843518
- [ ] 800727
- [ ] 254407
- [ ] 783919
- [ ] 571423
- [ ] 425248
- [ ] 245575

Puzzle 72

```
4 2 0 1 0 3 2 5 7 1 1 4 6 0 2 6
7 6 6 8 0 6 9 0 8 3 3 0 2 9 7 0
9 3 9 7 0 7 8 8 3 2 3 2 2 3 1 4
1 0 9 1 9 8 1 4 6 0 1 2 8 1 1 0
2 1 7 7 6 0 4 2 4 7 8 8 1 3 5 8
3 8 0 7 4 1 0 4 4 4 6 6 4 5 6 8
8 3 1 8 5 3 9 3 2 1 2 6 4 0 2 0
1 5 7 6 9 0 7 8 9 9 5 3 2 2 2 9
0 7 9 2 6 4 2 9 7 5 5 2 9 6 4 1
6 5 2 2 6 6 9 2 0 8 2 6 7 1 7 6
8 7 9 4 4 7 9 8 5 9 9 9 5 2 0 9
0 9 6 4 3 7 1 0 1 6 2 8 3 7 4 0
3 4 2 5 6 6 8 1 2 1 4 3 4 7 6 9
6 2 6 9 1 7 6 2 7 7 2 1 4 9 0 3
9 6 5 3 1 8 2 1 2 0 9 3 6 0 5 9
3 1 5 4 6 3 6 7 8 4 4 4 7 3 7 4
9 9 8 2 7 4 6 5 3 1 5 4 9 8 9 6
3 8 4 3 2 4 5 4 1 0 7 9 6 9 9 4
```

☐ 463678   ☐ 757538   ☐ 427251   ☐ 069083
☐ 397437   ☐ 479859   ☐ 842438   ☐ 564728
☐ 671962   ☐ 053139   ☐ 261017   ☐ 295657
☐ 668121   ☐ 792462   ☐ 622814   ☐ 742265

Puzzle 73

```
5 8 8 2 8 7 6 5 4 0 9 9 1 6 6 4
2 2 0 7 7 0 0 6 9 3 2 8 3 9 9
3 1 7 3 6 9 0 3 1 4 7 3 2 0 4 4
3 5 2 5 6 5 4 3 1 5 9 5 5 3 6 8
6 7 3 4 4 6 3 8 1 0 2 2 4 9 2 5
6 6 7 7 4 2 7 7 4 8 1 9 2 6 5 0
2 6 6 2 7 4 2 1 8 1 8 4 0 5 5 7
0 7 4 6 6 7 5 9 8 5 0 7 7 4 8 0
0 9 5 5 7 1 6 2 4 7 2 7 7 5 2 6
9 1 7 2 6 9 4 4 3 4 2 7 4 6 9 6
9 8 3 7 0 8 4 1 6 1 0 3 9 7 3 1
6 0 3 5 7 8 6 4 9 1 2 7 8 7 9 8
5 4 8 0 3 0 4 6 2 2 9 5 1 3 8 4
6 9 1 8 5 3 2 9 6 6 2 7 7 9 2 4
4 5 0 8 0 2 6 3 7 8 8 2 6 9 0 5
2 0 8 6 2 8 1 8 9 3 5 0 5 4 8 3
2 1 7 6 9 6 9 2 3 4 5 8 6 2 6 4
6 9 9 9 6 6 4 6 7 9 2 7 7 0 8 8 8
```

- ☐ 725644
- ☐ 764573
- ☐ 866686
- ☐ 855264
- ☐ 220183
- ☐ 947702
- ☐ 354726
- ☐ 461976
- ☐ 770069
- ☐ 545677
- ☐ 900266
- ☐ 192225
- ☐ 031473
- ☐ 671872
- ☐ 182680
- ☐ 176969

Puzzle 74

```
8 6 5 1 1 2 2 2 7 2 3 0 6 8 2 9
9 0 1 6 5 9 4 0 4 5 6 5 3 5 5 9
6 9 6 1 5 4 7 0 8 9 4 8 6 1 3 2
7 8 6 4 2 1 8 1 2 3 7 6 8 6 5 1
8 1 7 4 0 2 9 0 4 0 4 8 8 6 4 7
5 0 4 4 5 4 0 8 2 0 9 9 3 2 3 6
0 5 4 7 9 1 1 1 6 9 3 6 3 9 2 6
5 7 3 0 4 0 3 8 2 1 4 1 4 0 6 3
9 9 5 6 5 4 5 2 0 2 7 8 4 3 8 6
3 9 4 3 8 5 5 9 4 4 1 9 8 0 0 8
3 7 5 6 7 7 1 7 7 7 3 3 5 5 8 7
8 2 7 9 6 3 5 6 8 1 3 2 0 3 3 4
9 9 3 5 4 3 6 7 0 9 4 0 2 6 3 8
5 0 4 5 8 1 4 9 9 2 7 2 8 9 5 1
5 7 0 3 2 9 8 2 8 6 8 3 9 5 3 1
8 2 2 4 4 2 9 5 7 3 5 2 1 8 2 8
4 8 1 9 0 1 1 9 4 7 5 2 1 9 0 3
5 8 1 1 6 2 9 6 4 0 7 5 0 0 4 2
```

- ☐ 242620
- ☐ 181246
- ☐ 445408
- ☐ 208349
- ☐ 892417
- ☐ 798754
- ☐ 304037
- ☐ 537369
- ☐ 149599
- ☐ 547911
- ☐ 588492
- ☐ 553565
- ☐ 808623
- ☐ 335971
- ☐ 764824
- ☐ 896785

Puzzle 75

```
2 1 4 6 9 0 7 5 4 2 3 7 2 4 2 1
8 2 3 5 8 2 5 1 4 5 6 0 9 4 6 8
1 2 2 8 7 6 4 6 2 7 4 6 4 2 0 7
3 3 9 2 0 6 8 6 1 6 5 4 5 9 5 4
0 8 5 6 0 1 0 0 9 6 0 2 6 9 6 7
7 7 4 5 3 4 1 4 5 9 1 5 2 0 2 5
0 7 7 0 2 2 4 1 4 0 7 8 8 0 5 7
1 5 2 5 7 3 1 4 4 5 0 9 9 9 1 5
8 2 8 3 8 1 1 6 6 2 0 4 3 3 3 0
7 2 5 7 3 5 9 5 7 4 1 4 8 0 0 3
0 5 5 7 3 6 9 4 9 2 3 5 1 7 0 7
6 7 7 0 5 0 7 1 2 9 5 6 1 8 8 2
1 6 0 7 6 7 5 5 4 0 6 9 4 9 6 1
0 2 2 2 6 7 2 3 5 3 9 0 3 5 1 0
0 9 8 5 0 9 5 4 9 1 5 4 8 9 4 2
1 5 6 0 3 7 9 5 2 0 1 4 0 4 6 0
5 4 9 3 7 4 4 0 9 7 3 6 7 6 1 1
7 9 0 7 8 7 5 9 8 9 8 4 9 3 6 8
```

- ☐ 415345
- ☐ 735957
- ☐ 427324
- ☐ 083411
- ☐ 285328
- ☐ 444022
- ☐ 827459
- ☐ 644591
- ☐ 682075
- ☐ 607810
- ☐ 126058
- ☐ 999054
- ☐ 440965
- ☐ 429900
- ☐ 963216
- ☐ 469075

Puzzle 76

```
1 1 2 1 0 6 4 8 0 1 1 4 0 7 5 4
3 9 0 5 4 6 2 0 2 3 1 3 6 6 7 6
6 3 0 6 3 5 1 4 5 3 5 5 2 7 6 8
0 7 7 7 7 6 5 4 7 7 9 7 8 2 8 8
6 0 1 1 3 5 8 0 1 9 6 3 5 2 1 1
4 5 6 2 6 3 4 4 2 2 5 4 8 8 5 9
1 2 3 3 1 9 5 4 3 4 9 1 1 6 5 1
0 3 7 3 8 0 4 2 6 4 0 9 4 9 3 6
3 7 0 8 5 4 7 4 3 1 9 1 4 7 4 4
1 7 3 7 3 0 7 3 4 4 2 8 0 2 1 8
9 4 8 4 4 4 6 6 5 8 8 0 6 2 8 2
1 6 4 7 0 9 9 1 7 7 0 6 2 9 1 9
2 8 2 4 0 0 9 8 5 3 0 0 8 2 4 0
9 6 4 4 5 2 0 6 1 7 7 4 1 5 3 6
3 6 2 8 7 3 6 1 3 4 0 3 4 7 1 1
6 5 6 3 5 3 6 7 8 2 9 8 4 8 2 9
4 4 2 0 2 7 4 6 0 8 5 6 4 5 7 8
6 1 4 1 5 4 5 1 4 9 8 6 9 7 9 5
```

- [ ] 720244
- [ ] 445760
- [ ] 961733
- [ ] 437037
- [ ] 673734
- [ ] 570448
- [ ] 689434
- [ ] 044185
- [ ] 191293
- [ ] 085474
- [ ] 842426
- [ ] 253370
- [ ] 181431
- [ ] 846012
- [ ] 551867
- [ ] 429733

Puzzle 77

```
3 3 5 2 6 0 6 7 0 0 5 7 4 8 9 1
0 8 5 5 6 7 1 5 5 5 0 7 8 8 8 5
8 3 7 0 2 6 5 0 6 6 3 9 4 9 9 5
5 5 0 6 9 3 5 1 8 0 0 3 0 4 1 7
3 9 9 4 6 7 0 1 5 0 1 7 5 8 7 0
2 9 1 7 9 1 1 0 3 9 1 8 4 0 1 0
8 4 4 7 6 2 0 1 2 6 6 3 5 0 8 4
2 0 2 9 5 4 2 0 2 7 7 2 3 8 7 5
2 1 4 9 5 8 1 0 2 7 1 5 4 8 7 4
1 5 2 6 7 3 3 6 3 9 3 6 4 6 6 3
6 1 0 2 5 2 4 9 4 1 2 2 8 7 2 4
8 6 0 1 0 4 2 7 2 3 3 7 6 8 2 0
0 5 8 0 9 5 5 5 5 2 2 5 7 4 1 7
6 0 5 4 2 1 0 4 6 3 9 8 9 1 5 1
1 9 4 1 0 8 8 9 9 8 0 0 1 5 3 5
5 5 6 8 7 1 9 6 0 8 9 8 7 0 4 7
6 9 6 1 1 0 0 5 3 9 0 3 0 6 0 6
2 8 7 4 3 8 6 3 9 7 7 2 3 6 0 5
```

- ☐ 965243
- ☐ 419075
- ☐ 090923
- ☐ 590561
- ☐ 214958
- ☐ 606253
- ☐ 008867
- ☐ 853222
- ☐ 197684
- ☐ 301167
- ☐ 883231
- ☐ 010551
- ☐ 006975
- ☐ 888705
- ☐ 769006
- ☐ 110053

Puzzle 78

```
8 0 0 5 9 9 2 7 5 1 6 5 9 8 0 9
2 3 8 8 4 2 6 6 3 5 1 7 9 0 1 2
4 7 6 3 1 6 4 7 0 9 6 2 9 0 6 9
8 3 4 5 7 8 8 3 6 2 5 1 7 3 7 5
1 4 6 2 6 6 2 7 9 0 5 6 5 2 1 7
7 5 9 5 0 4 2 4 1 5 2 3 3 3 8 0
0 6 5 1 2 8 1 9 3 2 8 0 1 8 7 8
0 7 5 0 5 1 7 8 8 1 4 3 1 6 1 8
5 1 4 6 2 8 1 9 3 3 0 5 6 8 9 2
4 2 1 6 5 1 2 9 2 3 0 8 9 9 1 4
4 9 2 9 6 1 0 5 1 6 5 0 6 2 8 3
9 9 1 1 2 3 6 5 4 1 5 8 6 3 4 0
6 5 1 2 0 6 0 2 6 5 7 1 6 6 6 9
8 4 3 2 2 7 4 6 5 9 2 6 3 2 0 2
7 4 9 5 4 2 3 2 8 6 1 4 8 3 4 2
5 2 3 4 0 1 1 1 5 1 3 7 9 5 3 6
7 3 1 1 9 4 6 4 6 2 3 5 4 4 0 0
5 7 7 3 9 2 3 3 5 2 0 8 9 4 5 6
```

- ☐ 734567
- ☐ 142194
- ☐ 788362
- ☐ 123831
- ☐ 676020
- ☐ 248832
- ☐ 256202
- ☐ 865033
- ☐ 926425
- ☐ 735165
- ☐ 406481
- ☐ 689236
- ☐ 316470
- ☐ 005992
- ☐ 921561
- ☐ 746592

Puzzle 79

```
7 3 3 0 7 6 0 1 8 3 2 0 1 9 7 4
4 7 2 5 6 8 0 0 8 7 8 1 0 8 9 2
0 1 6 6 0 9 5 9 7 2 3 1 5 3 8 4
4 9 3 6 4 1 3 9 8 5 9 1 5 7 0 5
1 9 6 0 7 8 2 5 1 4 2 7 0 9 7 4
5 4 4 3 5 2 6 6 1 1 1 2 3 4 5 8
5 0 7 4 0 2 5 3 2 2 3 7 0 1 3 2
3 4 5 3 9 6 6 1 3 6 1 5 0 5 0 7
7 4 7 2 9 7 9 4 1 0 2 4 0 1 8 9
3 2 6 0 0 2 0 5 6 4 6 3 9 0 6 7
4 6 0 2 9 2 1 9 0 8 8 4 7 7 9 8
4 2 5 3 9 2 1 5 2 1 6 3 1 5 1 8
1 1 7 6 1 3 3 3 3 0 2 9 3 7 0 2
4 3 6 7 8 0 7 1 6 3 0 9 8 0 7 1
9 9 0 4 5 3 5 7 5 7 0 3 0 4 0 9
4 0 5 7 0 8 4 4 4 4 4 1 2 5 9 8
1 0 5 2 3 1 8 7 5 9 0 4 7 7 4 8
8 0 1 1 1 5 5 4 4 0 5 7 3 8 7 6
```

- [ ] 739215
- [ ] 499173
- [ ] 480750
- [ ] 411527
- [ ] 477409
- [ ] 293774
- [ ] 780086
- [ ] 092315
- [ ] 330293
- [ ] 326262
- [ ] 131293
- [ ] 018320
- [ ] 505202
- [ ] 070196
- [ ] 748809
- [ ] 510757

Puzzle 80

```
0 1 4 0 0 3 2 3 7 3 0 7 2 9 3 0
9 0 4 2 7 8 8 4 0 9 9 6 3 4 0 2
8 3 3 8 2 7 1 2 8 8 4 3 3 0 5 5
9 6 1 2 4 3 1 4 8 6 5 7 4 0 5 2
1 7 6 7 1 7 6 2 0 1 2 4 7 3 6 4
9 1 7 9 5 7 5 8 2 6 6 4 1 4 0 6
7 4 7 1 7 5 7 2 1 2 6 5 1 2 7 8
2 9 6 4 1 8 2 3 5 8 6 2 9 3 3 1
8 3 1 8 3 7 8 2 6 8 1 4 4 0 3 2
0 0 0 9 0 6 3 5 9 4 8 7 6 4 0 0
1 9 9 1 4 6 0 3 3 8 1 0 9 0 5 2
9 8 0 8 0 4 6 6 2 4 8 4 9 9 0 5
1 3 0 6 1 1 3 7 8 9 3 3 1 4 1 6
8 1 1 9 0 1 5 7 4 0 3 9 9 1 2 5
5 1 4 4 5 0 1 2 3 7 9 5 5 5 8 9
4 4 1 3 9 2 2 8 3 6 3 0 1 7 3 0
0 9 1 2 3 3 9 0 0 5 0 1 1 1 1 4
4 0 9 7 3 1 9 5 7 4 8 7 6 5 7 4
```

- ☐ 950104
- ☐ 012379
- ☐ 271288
- ☐ 117332
- ☐ 893314
- ☐ 900141
- ☐ 287381
- ☐ 191854
- ☐ 523282
- ☐ 442660
- ☐ 005011
- ☐ 501283
- ☐ 417630
- ☐ 072415
- ☐ 904887
- ☐ 034393

Puzzle 81

```
8 3 3 6 9 3 5 7 8 7 8 4 3 0 6 8
1 7 7 4 3 9 9 1 8 0 8 6 5 1 9 6
9 4 0 0 4 2 4 9 7 4 2 6 5 1 8 8
0 9 4 7 1 5 6 2 7 3 3 9 9 0 7 5
0 9 0 4 0 9 0 5 8 3 1 8 1 2 6 5
3 3 9 7 9 8 8 1 8 1 7 5 2 1 7 7
9 9 5 0 5 0 8 6 5 9 1 2 0 3 2 5
0 9 1 6 0 8 9 8 9 0 2 7 7 5 7 6
8 0 3 4 4 2 0 6 3 9 6 8 9 3 1 3
5 4 8 9 0 8 0 9 3 9 0 5 3 1 6 6
0 0 9 2 3 2 3 6 3 0 4 1 8 1 0 6
0 5 9 7 1 0 1 4 6 8 3 9 5 1 5 6
6 8 7 3 3 6 2 6 6 9 2 4 3 5 1 8
0 6 7 2 3 4 2 2 2 0 0 7 8 9 3 2
5 3 6 6 9 3 1 6 3 7 2 6 9 8 0 9
3 9 9 5 9 6 6 9 8 1 1 8 2 6 8 7
4 1 6 1 9 8 1 8 6 7 9 4 2 9 3 3
0 1 8 0 8 8 3 8 1 1 7 8 6 1 4 8
```

- ☐ 683266
- ☐ 050599
- ☐ 199347
- ☐ 881817
- ☐ 276789
- ☐ 987002
- ☐ 118560
- ☐ 583181
- ☐ 839493
- ☐ 821626
- ☐ 878430
- ☐ 699015
- ☐ 808651
- ☐ 430369
- ☐ 265174
- ☐ 809382

Puzzle 82

```
2 0 0 3 1 8 4 0 1 1 1 0 1 2 2 0
0 8 9 3 5 6 1 6 1 9 4 9 6 0 9 9
2 4 0 5 5 6 1 0 8 5 8 1 5 3 1 0
7 2 0 0 9 4 0 8 9 6 5 7 7 6 4 6
6 2 9 5 2 3 9 3 0 2 1 2 9 2 0 7
1 0 1 3 0 9 4 2 1 3 3 4 9 4 8 0
8 1 2 6 4 0 5 9 5 9 5 2 7 2 8 7
8 4 5 8 4 4 1 3 7 7 7 0 4 4 8 8
0 1 5 4 0 7 3 9 9 1 0 4 7 2 0 3
2 6 8 0 5 4 1 3 2 4 8 1 7 0 8 7
3 0 2 8 8 3 4 5 9 4 1 0 7 9 3 9
3 2 2 6 6 4 6 8 8 3 1 5 5 7 3 6
3 5 4 1 1 7 2 0 1 9 4 4 4 6 1 1
2 3 9 1 1 9 3 8 3 7 3 9 7 1 5 1
8 7 2 7 2 3 7 3 0 7 6 8 7 6 3 2
7 1 2 6 6 6 6 0 5 9 9 4 6 8 3 1
4 8 2 4 2 4 1 7 8 6 4 6 8 9 8 1
3 7 6 0 1 4 0 9 5 1 9 5 8 1 2 9
```

- [ ] 349718
- [ ] 635053
- [ ] 091255
- [ ] 302883
- [ ] 463974
- [ ] 580165
- [ ] 916165
- [ ] 203624
- [ ] 864689
- [ ] 302129
- [ ] 808880
- [ ] 861679
- [ ] 394154
- [ ] 001924
- [ ] 823332
- [ ] 680541

Puzzle 83

```
2 0 1 9 5 3 9 0 4 7 7 9 9 4 6 6
2 8 5 8 9 2 3 1 1 0 7 5 1 9 1 5
1 2 2 1 7 6 0 4 4 3 7 5 7 5 5 5
6 4 9 3 1 8 9 3 0 5 1 4 4 2 7 2
0 3 6 0 4 0 6 8 8 1 1 3 8 4 5 2
6 3 3 4 4 5 6 8 8 8 7 0 9 4 2 9
4 3 0 9 3 6 8 4 0 8 1 4 6 7 9 4
8 2 1 2 5 5 5 4 4 0 6 5 2 1 6 1
6 8 2 3 3 5 0 3 6 6 6 7 5 8 6 7
9 6 1 0 4 9 0 1 8 8 0 9 0 2 2 7
8 2 7 6 0 8 6 3 5 7 3 7 5 6 4 4
9 6 9 1 1 6 6 5 5 6 7 0 3 3 9 8
1 4 6 5 3 9 2 0 7 7 3 9 7 3 0 8
3 1 0 0 1 0 5 8 9 1 9 4 7 3 2 5
9 5 8 9 9 2 3 0 7 2 0 5 9 3 8 6
4 0 9 4 7 0 8 1 7 4 3 8 0 4 0 9
8 1 0 7 3 7 2 0 1 1 6 7 8 3 0 2
9 6 9 3 7 3 9 7 8 5 0 7 9 3 2 9
```

- ☐ 090668
- ☐ 049018
- ☐ 923110
- ☐ 881138
- ☐ 125604
- ☐ 596008
- ☐ 093591
- ☐ 630121
- ☐ 249078
- ☐ 375368
- ☐ 705916
- ☐ 490280
- ☐ 606486
- ☐ 743804
- ☐ 406712
- ☐ 773973

Puzzle 84

```
7 8 3 4 9 7 3 2 0 5 7 2 9 8 0 7
7 3 4 9 1 8 9 6 1 7 6 6 1 4 4 4
9 2 0 4 6 5 1 3 2 5 8 2 1 1 7 6
2 0 3 6 8 2 8 7 4 4 7 2 5 0 4 5
6 4 1 3 1 6 1 5 0 9 2 0 2 9 8 7
9 9 9 2 6 9 7 9 4 1 2 6 2 0 6 4
0 7 5 1 3 1 6 1 2 3 2 2 5 2 7 5
6 3 7 7 9 5 2 9 1 2 5 0 5 9 5 1
6 5 4 5 8 6 9 6 8 0 0 6 5 1 8 3
2 7 4 4 8 6 9 0 6 9 1 5 1 8 5 9
1 6 1 6 8 2 1 2 0 3 4 4 7 6 9 6
4 0 9 9 9 4 9 6 3 0 5 4 6 6 6 9
3 3 9 1 3 3 6 3 1 3 5 6 3 1 0 1
5 1 1 3 9 8 4 6 5 6 9 0 3 7 8 1
6 7 9 7 2 2 4 8 6 9 4 6 9 5 8 6
0 8 8 9 0 4 8 9 3 7 4 5 0 3 2 7
6 9 8 3 1 2 4 4 1 8 9 5 7 4 3 9
1 3 4 5 8 8 4 5 6 2 2 1 4 8 2 2
```

- [ ] 181762
- [ ] 984098
- [ ] 766648
- [ ] 719384
- [ ] 645712
- [ ] 906621
- [ ] 768474
- [ ] 923396
- [ ] 829359
- [ ] 541052
- [ ] 918661
- [ ] 616875
- [ ] 262149
- [ ] 918961
- [ ] 691928
- [ ] 202905

85

Puzzle 85

```
2 9 4 1 1 9 1 5 8 5 0 2 6 7 7 5
4 6 1 9 5 3 2 0 5 0 7 5 3 2 6 8
6 1 3 6 6 4 2 5 1 5 3 8 7 2 8 6
2 4 7 6 1 2 0 5 1 9 5 9 7 6 6 5
3 0 7 8 6 7 8 6 9 8 1 6 5 3 4 5
1 9 3 8 1 7 1 9 8 7 1 1 0 7 6 8
2 1 6 6 7 2 5 4 4 3 6 0 1 3 3 2
5 4 0 3 0 8 8 7 0 2 3 5 8 5 2 1
0 5 5 8 1 7 8 6 7 7 6 8 1 9 2 4
8 4 1 0 6 7 5 8 7 5 4 1 3 7 8 8
0 1 9 1 5 3 9 9 8 4 8 4 3 5 5 2
8 4 4 0 7 5 7 7 0 8 5 2 8 5 0 2
1 7 8 9 6 1 3 3 4 8 1 0 1 1 8 4
9 1 7 3 0 9 2 9 0 4 0 0 8 0 7 5
3 8 9 4 7 3 4 7 0 5 8 9 8 3 4 7
2 2 7 5 8 2 6 1 9 5 0 1 8 2 3 6
7 6 5 9 5 3 6 7 6 9 0 5 8 3 8 0
6 5 8 6 3 7 6 5 3 5 3 7 8 0 6 4
```

- ☐ 101911
- ☐ 704891
- ☐ 515246
- ☐ 368011
- ☐ 789613
- ☐ 101843
- ☐ 502167
- ☐ 166725
- ☐ 850967
- ☐ 875413
- ☐ 903719
- ☐ 171839
- ☐ 235916
- ☐ 810591
- ☐ 708004
- ☐ 915399

Puzzle 86

```
4 2 2 4 9 7 5 9 7 1 3 5 4 5 9 3
8 9 6 8 3 5 4 9 9 6 3 9 9 3 0 4
0 3 8 4 1 4 2 6 1 0 1 2 2 7 9 3
0 3 4 4 7 9 6 2 6 6 0 2 2 7 0 5
7 2 4 1 4 2 5 0 9 4 3 3 1 4 7 8
2 2 0 3 4 6 5 3 8 7 3 0 0 0 4 2
6 8 9 9 9 5 4 9 2 8 3 9 2 3 8 9
8 9 0 4 9 7 9 9 1 8 0 7 2 8 7 9
0 2 7 2 4 8 8 3 2 6 9 8 2 0 6 3
5 7 4 0 0 0 7 7 3 4 4 4 5 7 1 1
9 7 4 3 1 3 9 7 2 4 2 8 5 2 3 3
8 4 6 2 4 4 4 4 3 8 2 8 7 7 7 7
9 1 1 6 0 5 7 7 4 9 8 5 3 4 4 4
1 2 3 6 1 4 4 5 9 6 2 5 6 3 0 3
0 7 2 0 8 0 8 2 1 7 6 8 0 4 5 0
3 6 7 6 5 3 8 6 1 1 5 8 9 2 1 8
6 6 0 9 8 3 4 5 0 8 0 4 8 0 1 4
8 0 7 4 9 7 1 4 3 6 8 0 1 5 3 2
```

- [ ] 739930
- [ ] 269544
- [ ] 313972
- [ ] 345080
- [ ] 990030
- [ ] 611589
- [ ] 644894
- [ ] 916487
- [ ] 808217
- [ ] 737922
- [ ] 137405
- [ ] 960662
- [ ] 927741
- [ ] 044927
- [ ] 280203
- [ ] 341459

Puzzle 87

```
7 2 1 5 0 4 7 9 3 1 3 9 0 6 4 3
8 3 3 7 8 0 2 8 6 1 1 3 5 5 9 4
7 2 6 4 2 2 1 7 2 8 5 9 3 3 6 1
7 4 1 4 6 8 8 9 1 6 1 1 1 1 1 3
5 4 7 3 7 3 9 5 7 8 2 4 8 4 8 2
1 6 5 4 6 3 2 4 5 0 7 5 0 9 1 5
0 9 1 4 9 9 4 5 6 9 3 2 0 6 1 8
9 8 2 9 3 9 2 8 8 7 5 8 9 6 6 1
1 5 5 2 2 7 6 1 3 0 6 6 6 9 1 2
9 8 8 6 5 2 4 1 6 4 0 3 8 7 3 4
9 6 8 3 8 1 7 5 5 2 7 3 1 0 4 0
8 9 9 8 5 9 3 8 5 7 7 7 7 3 6 4
2 4 0 9 8 9 6 3 7 3 5 3 1 5 6 0
0 5 0 2 2 7 8 5 0 4 1 7 6 2 6 9
0 8 1 8 9 1 6 1 2 7 5 3 8 7 7 8
0 1 1 0 2 1 7 3 4 9 1 4 3 2 1 8
2 2 2 1 9 0 3 1 3 8 2 5 4 4 0 9
6 7 6 2 1 3 4 1 0 8 8 0 5 9 7 9
```

- ☐ 452831
- ☐ 877510
- ☐ 028611
- ☐ 678888
- ☐ 372612
- ☐ 149669
- ☐ 371201
- ☐ 054236
- ☐ 091222
- ☐ 158351
- ☐ 697035
- ☐ 737395
- ☐ 577065
- ☐ 446985
- ☐ 997219
- ☐ 298362

Puzzle 88

```
8 6 2 3 6 7 6 1 3 1 2 1 4 1 4 2
9 6 1 9 2 7 8 3 7 7 6 0 5 4 1 4
3 2 9 9 9 2 6 6 4 1 7 7 2 4 8 8
1 4 5 0 9 1 1 0 0 1 5 4 5 3 3 8
0 7 8 2 0 2 0 2 0 0 6 0 1 2 7 9
2 1 3 1 4 0 5 7 4 3 6 8 4 4 3 5
8 6 0 8 1 1 4 7 2 6 2 3 9 3 1 9
1 2 9 5 9 7 8 3 8 4 3 9 4 6 1 1
8 8 1 0 0 3 5 2 4 5 6 3 8 9 1 3
3 0 1 5 2 4 9 6 9 2 1 1 2 6 3 7
5 6 5 7 7 7 9 4 3 0 6 5 5 1 0 1
1 5 1 8 5 1 7 3 2 3 5 1 4 8 9 1
5 7 7 5 6 3 2 4 1 3 3 8 9 6 7 2
3 5 0 0 6 8 4 6 1 9 7 2 3 3 7 2
6 0 6 4 0 7 3 2 9 3 4 4 4 3 4 3
1 0 2 3 9 5 5 1 3 4 4 5 3 8 4 0
3 0 2 3 9 9 9 1 7 2 1 0 0 3 3 7
7 7 2 6 2 6 3 7 6 3 1 0 7 8 0 7
```

- ☐ 102395
- ☐ 732643
- ☐ 251494
- ☐ 821571
- ☐ 314236
- ☐ 102127
- ☐ 121316
- ☐ 940501
- ☐ 198365
- ☐ 732351
- ☐ 648600
- ☐ 306551
- ☐ 351536
- ☐ 863475
- ☐ 544315
- ☐ 990419

Puzzle 89

```
3 9 9 8 7 0 4 9 9 4 6 5 2 2 2 5
7 0 6 4 8 0 4 0 2 7 4 1 1 8 2 1
1 9 1 9 4 5 9 3 6 6 6 4 1 8 8 0
2 9 1 8 9 9 2 2 4 6 7 3 1 9 8 8
6 6 8 0 7 9 6 7 0 6 4 8 2 0 3 1
3 6 5 2 7 0 3 6 1 6 0 9 5 1 1 0
2 3 6 8 0 7 9 6 6 9 7 9 6 7 5 4
0 1 2 0 8 8 2 1 4 3 0 1 3 1 9 7
3 7 1 1 1 2 7 2 9 8 7 8 9 8 6 4
9 6 8 9 8 7 2 1 5 5 9 5 2 0 4 9
3 7 7 1 8 3 7 4 6 7 3 7 6 0 8 9
9 9 8 9 1 3 7 2 0 3 2 6 4 4 5 7
6 9 6 7 3 5 2 8 1 3 7 3 1 3 6 5
6 4 4 0 7 6 5 3 2 9 7 1 6 2 6 1
4 0 4 3 3 6 7 0 0 1 2 3 0 4 3 3
4 0 4 6 1 4 7 8 6 1 0 4 4 6 0 8
6 1 8 0 1 5 9 3 5 2 6 8 6 8 7 6
7 1 9 7 5 2 2 4 0 6 5 4 5 2 5 9
```

- ☐ 159648
- ☐ 163710
- ☐ 979669
- ☐ 987049
- ☐ 125639
- ☐ 422885
- ☐ 533728
- ☐ 801287
- ☐ 446878
- ☐ 465736
- ☐ 396644
- ☐ 590616
- ☐ 945936
- ☐ 607697
- ☐ 437304
- ☐ 718004

Puzzle 90

```
9 7 1 7 6 7 0 7 6 3 9 6 1 0 7 8
5 0 2 8 1 5 8 0 9 6 6 1 9 3 1 3
3 3 7 7 4 1 1 0 9 8 6 5 2 7 0 7
2 8 6 4 0 4 4 0 6 4 6 3 7 7 8 5
1 2 5 8 0 1 4 1 6 1 4 5 5 9 4 7
1 8 0 1 5 2 0 8 9 9 4 6 8 2 0 0
9 3 4 4 4 6 4 0 1 3 7 1 7 2 0 1
5 3 9 5 6 9 7 9 0 2 0 6 8 6 6 3
7 1 9 9 7 8 3 1 3 3 4 2 7 3 1 7
6 3 0 2 7 8 8 8 2 5 6 3 4 6 2 1
7 9 9 7 3 2 4 1 0 2 7 6 2 2 4 3
4 8 8 1 5 9 1 3 4 4 4 2 2 5 6 7
5 2 5 7 4 5 4 6 8 5 0 1 9 7 7 6
5 5 3 5 4 3 3 6 7 4 1 0 4 4 5 8
7 2 2 6 7 1 7 6 8 6 6 8 5 3 1 1
7 4 8 4 5 0 5 4 3 8 7 0 4 0 9 2
2 3 1 3 1 9 9 3 2 7 1 9 2 0 2 6
0 7 0 9 8 0 3 1 9 9 3 4 2 5 2 5
```

☐ 997216	☐ 785570	☐ 440646	☐ 860209
☐ 139166	☐ 473419	☐ 804005	☐ 639610
☐ 792263	☐ 056721	☐ 483740	☐ 541614
☐ 612467	☐ 126988	☐ 392394	☐ 899468

Puzzle 91

```
0 5 7 8 4 5 3 1 3 0 1 9 9 9 7 7
6 1 1 2 8 7 8 6 7 3 3 0 8 7 8 6
6 3 0 1 9 5 6 1 1 4 3 1 3 7 8 1
0 8 5 4 4 0 7 7 4 7 5 3 6 3 2 6
1 3 9 1 8 0 4 3 5 6 4 4 7 9 4 9
1 2 1 9 7 4 0 4 1 1 4 4 8 4 7 1
5 8 1 4 8 8 1 9 5 9 6 2 3 7 8 2
6 4 3 4 7 3 2 9 8 9 3 6 1 2 5 9
6 2 7 0 5 3 8 2 3 2 8 0 9 6 5 4
9 2 6 7 6 2 7 3 8 1 9 1 0 1 0 0
0 2 0 8 1 4 3 8 5 4 4 1 0 9 9 2
1 4 0 0 9 9 5 4 6 3 2 5 1 3 9 4
9 6 8 1 8 8 4 9 6 8 9 5 2 1 7 6
8 5 9 8 2 0 1 6 4 9 7 7 0 8 9 6
3 7 8 1 3 3 4 2 1 5 6 8 9 2 8 1
0 3 4 5 3 9 3 2 8 2 6 7 4 7 8 0
2 6 8 8 4 1 4 1 6 0 3 5 3 7 3 6
7 0 7 6 8 9 0 2 9 4 7 6 5 5 9 0
```

- [ ] 523469
- [ ] 034761
- [ ] 635178
- [ ] 378210
- [ ] 829111
- [ ] 619823
- [ ] 173499
- [ ] 162749
- [ ] 826747
- [ ] 061414
- [ ] 901344
- [ ] 990039
- [ ] 882541
- [ ] 115669
- [ ] 678319
- [ ] 422246

Puzzle 92

```
8 5 0 4 1 3 7 9 6 0 9 5 0 2 3 2
8 3 2 7 9 1 6 5 2 2 4 1 5 6 8 3
6 2 0 6 3 8 4 8 9 4 7 8 9 2 9 4
9 9 7 0 8 8 2 7 4 3 7 7 6 3 6 3
9 1 2 7 7 7 5 9 3 4 9 2 7 2 9 3
2 3 8 0 2 0 0 2 7 2 5 6 1 3 9 2
5 7 0 6 1 9 2 1 8 9 2 0 4 5 9 3
1 3 5 1 6 7 7 5 0 3 5 1 3 2 7 2
7 4 2 4 5 6 4 7 4 6 8 9 8 7 8 1
0 7 3 7 8 0 7 1 9 4 2 7 0 7 2 8
8 6 0 7 2 0 2 6 9 9 1 7 4 8 7 3
5 5 7 9 5 8 6 8 1 6 9 7 9 7 6 9
6 6 5 6 1 4 0 1 8 0 4 1 5 4 9 4
1 4 4 7 9 9 3 7 9 9 4 3 6 8 1 6
7 3 5 4 7 4 7 6 4 9 8 3 0 4 8 2
4 4 9 0 0 6 3 6 9 7 8 2 6 3 0 7
3 7 4 2 7 4 1 0 8 1 6 7 4 9 2 6
7 3 3 1 8 0 2 0 5 7 6 0 8 2 9 8
```

- ☐ 681766
- ☐ 379609
- ☐ 197771
- ☐ 224156
- ☐ 830070
- ☐ 478382
- ☐ 981994
- ☐ 472603
- ☐ 855776
- ☐ 725613
- ☐ 346567
- ☐ 561972
- ☐ 874843
- ☐ 799969
- ☐ 734728
- ☐ 294395

Puzzle 93

```
4 8 9 8 0 5 0 8 3 5 2 6 4 8 8 0
9 9 3 5 9 3 6 8 1 6 6 1 6 6 9 7
3 4 2 9 1 2 3 0 6 9 5 5 5 1 9 0
0 6 1 1 1 8 5 0 0 4 8 9 5 9 7 1
0 7 3 7 0 6 3 3 3 7 9 5 0 8 4 5
9 8 1 4 1 2 4 6 6 7 8 6 9 0 1 2
1 7 6 2 7 1 8 6 4 4 3 0 5 0 7 5
6 6 5 7 3 6 6 4 5 3 0 9 7 4 9 9
7 4 1 5 3 4 7 3 0 3 8 6 7 9 1 1
7 7 7 8 2 7 3 8 0 1 0 4 9 6 0 2
9 4 7 5 5 5 7 7 9 0 8 0 7 8 9 4
4 1 2 7 6 9 8 5 6 5 1 9 0 0 3 6
6 8 8 8 3 1 0 0 2 7 7 9 4 5 4 5
1 0 2 9 8 7 4 0 9 1 0 6 3 7 0 7
7 9 5 9 1 2 0 6 7 0 2 4 9 9 1 1
9 6 7 9 2 6 7 1 4 4 5 6 5 7 4 2
4 2 8 7 1 4 5 7 1 8 9 6 0 9 6 0
2 6 4 3 2 7 2 4 9 5 3 3 8 1 9 6
```

- ☐ 546637
- ☐ 839164
- ☐ 176271
- ☐ 544176
- ☐ 651900
- ☐ 655095
- ☐ 846253
- ☐ 688831
- ☐ 618639
- ☐ 793730
- ☐ 210284
- ☐ 552780
- ☐ 596032
- ☐ 801049
- ☐ 805089
- ☐ 097087

Puzzle 94

```
2 2 1 9 2 7 9 1 7 9 8 0 7 4 7 2
3 0 8 0 7 6 3 8 5 7 3 2 9 2 7 7
5 7 7 1 5 8 8 4 5 1 6 0 2 2 1 7
8 1 4 3 0 3 2 9 1 0 6 7 9 0 6 6
7 5 7 8 0 1 1 3 4 2 2 5 5 3 6 1
0 7 2 5 9 2 9 9 3 6 5 7 2 9 4 3
5 7 9 7 0 4 0 2 9 6 2 4 2 6 5 3
1 3 7 9 0 5 5 7 6 9 1 7 7 0 5 5
8 8 2 0 0 5 4 9 7 5 9 7 4 2 7 0
4 8 0 4 5 6 1 5 4 0 9 4 5 2 3 5
0 0 6 2 0 3 1 3 7 4 1 5 5 3 8 3
3 6 6 0 5 6 2 2 9 1 9 3 2 6 2 9
6 7 1 4 3 4 8 0 8 3 6 6 2 1 0 7
5 6 6 7 4 7 4 5 0 1 7 1 3 3 9 4
6 1 4 1 4 6 0 1 3 6 8 8 4 0 4 2
2 4 6 8 5 2 4 9 1 4 6 8 7 3 8 6
6 7 3 9 8 0 9 3 5 8 5 8 3 0 2 2
3 1 6 3 5 9 9 6 1 2 2 7 0 2 3 0
```

- [ ] 020770
- [ ] 632206
- [ ] 341254
- [ ] 747089
- [ ] 013688
- [ ] 595629
- [ ] 676088
- [ ] 945944
- [ ] 554661
- [ ] 836621
- [ ] 927917
- [ ] 931507
- [ ] 505457
- [ ] 554722
- [ ] 864194
- [ ] 143032

Puzzle 95

```
2 5 2 7 8 5 7 0 3 1 2 1 2 0 0 5
3 7 4 5 5 0 8 0 3 0 8 6 6 2 2 3
3 2 1 2 0 5 2 5 9 3 0 8 9 6 1 5
5 1 0 6 3 9 4 7 0 5 4 7 2 1 3 8
5 7 8 5 8 9 0 0 4 5 9 2 2 9 8 5
5 5 7 3 8 3 9 1 8 3 6 8 0 5 2 2
9 1 2 5 5 9 1 5 3 1 4 8 5 5 3 7
5 4 4 8 6 0 7 7 9 5 5 2 7 6 3 5
4 3 5 9 0 5 1 7 2 5 4 9 9 9 4 7
3 3 6 1 4 1 0 6 3 4 9 9 9 7 8 4
1 5 8 3 9 3 7 6 4 0 3 3 5 0 7 9
5 1 0 2 2 1 1 7 5 2 1 6 8 9 3 6
9 8 0 0 6 7 5 5 2 7 4 0 7 9 4 2
1 8 8 7 4 1 4 5 6 8 4 7 9 8 4 7
1 4 3 0 2 4 8 5 7 8 7 8 3 4 8 5
4 6 2 0 1 3 6 2 6 4 4 1 3 4 2 3
2 7 6 5 9 7 9 0 4 4 2 4 0 0 0 6
9 3 7 2 1 0 2 8 7 1 0 2 6 5 9 3
```

☐ 744139   ☐ 046739   ☐ 716831   ☐ 081599

☐ 823348   ☐ 725576   ☐ 462013   ☐ 544860

☐ 141633   ☐ 715095   ☐ 587252   ☐ 659790

☐ 907965   ☐ 599324   ☐ 865058   ☐ 434720

Puzzle 96

```
8 2 1 0 6 2 6 6 6 2 3 3 5 3 8 4 3
1 8 4 0 0 5 8 2 5 8 9 6 7 9 0 2
4 5 7 4 5 9 7 8 5 4 8 6 0 0 9 4
1 8 6 3 6 2 9 7 7 2 6 4 5 9 4 0
9 2 6 2 6 6 6 0 5 5 9 5 9 3 5 1
7 8 8 0 6 9 5 3 6 9 7 3 7 4 0 7
7 6 0 8 4 5 0 7 6 8 5 7 4 4 3 8
3 6 3 2 8 5 4 0 7 0 0 4 6 0 6 6
5 4 1 2 3 3 3 9 1 7 9 9 9 6 1 2
7 2 9 3 8 3 3 9 5 8 6 8 9 1 9 0
6 4 2 2 6 1 1 5 4 8 4 3 0 2 1 2
3 7 4 2 0 0 1 6 0 9 0 4 2 1 5 1
0 8 2 2 7 2 8 2 7 5 4 4 6 5 2 1
8 4 2 0 3 0 8 6 0 4 9 8 5 7 3 3
3 6 3 6 7 7 3 2 5 9 5 1 4 8 1 7
9 3 7 4 6 0 5 5 9 4 1 2 3 5 0 3
7 6 9 4 7 7 7 6 3 2 3 0 3 4 5 1
1 8 4 4 3 3 7 2 9 5 5 1 3 7 0 3
```

☐ 690018	☐ 548494	☐ 642261	☐ 362501
☐ 609900	☐ 367537	☐ 288494	☐ 894735
☐ 562625	☐ 913086	☐ 519163	☐ 690579
☐ 223228	☐ 858286	☐ 064739	☐ 331188

Puzzle 97

```
4 2 9 4 6 4 7 8 1 0 6 6 5 3 4 2
7 3 6 1 2 4 6 6 6 8 7 7 3 2 7 3
6 6 6 5 1 5 5 4 6 1 3 8 4 3 3 0
2 5 4 1 0 1 8 9 0 0 6 8 9 4 0 6
1 7 0 2 9 8 4 2 1 7 2 5 3 3 4 6
5 6 7 4 7 5 1 1 7 8 2 0 1 3 2 5
3 5 3 7 0 0 7 4 7 9 9 3 1 0 9 9
7 9 6 9 9 7 2 1 1 1 9 6 7 7 4 9
3 6 4 9 2 1 0 6 3 8 3 2 1 4 4 0
9 5 4 8 5 5 5 0 7 1 6 8 7 0 7 1
2 3 7 7 9 2 4 1 6 1 4 2 2 1 7 0
6 8 3 6 5 5 1 0 4 4 8 9 8 2 4 1
2 1 1 3 1 0 0 9 6 7 3 5 5 9 9 5
6 7 8 9 1 5 9 7 9 3 8 9 2 5 6 0
3 7 7 9 5 9 3 3 4 6 9 9 9 7 7 6
9 9 2 0 2 1 4 7 3 7 4 5 8 6 7 5
1 1 5 2 7 0 6 5 8 7 5 9 9 7 8 4
4 9 8 6 7 6 8 8 0 1 7 9 7 2 1 5
```

- ☐ 351267
- ☐ 485550
- ☐ 258176
- ☐ 874649
- ☐ 664216
- ☐ 343307
- ☐ 936262
- ☐ 701698
- ☐ 314955
- ☐ 412029
- ☐ 065875
- ☐ 508141
- ☐ 827992
- ☐ 979389
- ☐ 854737
- ☐ 040700

Puzzle 98

```
1 0 2 3 3 6 2 3 5 0 2 6 4 0 4 2
5 6 9 6 1 7 4 9 6 9 1 6 8 5 2 5
5 8 3 9 0 7 2 4 2 9 0 1 0 3 4 9
2 5 6 8 2 7 3 3 1 4 4 8 7 1 6 1
8 2 2 6 9 7 8 7 6 0 7 4 3 4 3 1
0 9 3 2 0 7 0 3 0 2 8 5 0 1 0 9
6 6 2 9 7 6 0 7 1 4 3 8 0 6 5 8
6 5 6 6 3 6 8 0 7 5 1 9 2 7 1 8
7 3 8 8 0 7 6 5 8 2 6 9 6 6 6 8
6 5 5 2 0 1 0 5 4 1 1 3 1 1 7 5
0 4 7 2 7 1 5 0 7 3 2 5 9 0 8 0
7 4 8 7 7 3 0 6 2 8 4 3 2 7 5 2
1 3 1 1 4 0 8 6 5 0 4 2 9 0 0 9
2 2 6 2 4 3 0 7 7 2 0 1 4 9 8 9
0 4 7 4 7 5 9 2 3 4 2 0 8 1 6 1
0 0 6 3 8 8 0 7 3 6 2 9 5 8 7 6
2 0 3 3 9 9 6 9 3 9 5 0 9 0 0 7
9 2 4 2 6 3 5 3 6 5 8 6 7 7 7 4
```

- ☐ 876074
- ☐ 303117
- ☐ 987447
- ☐ 258619
- ☐ 167610
- ☐ 535443
- ☐ 692689
- ☐ 587615
- ☐ 315639
- ☐ 105820
- ☐ 187586
- ☐ 860508
- ☐ 592342
- ☐ 993218
- ☐ 924056
- ☐ 404620

Puzzle 99

```
5 1 0 0 2 4 5 7 2 0 0 5 1 9 4 9
5 6 0 1 4 6 8 5 6 4 1 5 6 7 1 2
6 5 1 7 2 5 3 6 7 6 5 1 2 7 0 7
5 3 6 3 6 9 9 3 6 3 3 3 0 5 8 8
0 6 8 4 7 6 3 1 6 7 3 8 0 9 7 3
1 9 7 1 8 4 5 5 6 6 5 5 0 2 5 5
6 2 2 1 6 1 6 7 5 9 9 7 0 3 7 2
8 2 8 5 6 5 0 1 1 0 8 0 8 8 1 5
1 9 6 7 0 7 3 8 9 1 3 6 5 9 9 0
2 6 5 0 1 6 9 0 6 3 4 0 7 3 3 6
3 3 2 8 1 2 8 8 0 6 8 2 0 2 8 4
0 7 2 5 2 0 7 0 6 5 0 4 8 4 5 6
9 8 7 7 6 0 3 4 8 2 9 5 8 4 2 9
7 5 9 3 7 0 3 3 1 3 0 3 5 5 4 7
0 5 8 3 7 3 0 7 0 9 1 5 5 8 7 3
1 7 9 4 3 8 1 7 8 2 0 2 1 2 0 8
6 7 5 9 5 6 0 5 5 7 8 6 0 3 2 3
1 5 9 3 4 4 1 0 2 5 1 4 0 4 5 1
```

☐ 759970	☐ 787302	☐ 268976	☐ 564156
☐ 329577	☐ 847631	☐ 084389	☐ 651725
☐ 274190	☐ 369936	☐ 256007	☐ 253950
☐ 641065	☐ 754200	☐ 610562	☐ 794381

Puzzle 100

```
5 0 7 1 4 6 8 2 5 0 5 6 3 5 4 4
7 2 1 8 4 0 8 3 4 0 9 8 5 0 5 0
3 9 2 5 3 7 8 6 6 8 7 4 5 7 7 6
7 3 3 3 6 3 1 1 9 6 5 5 7 1 4 8
5 3 9 0 4 3 9 8 0 9 1 2 1 4 0 4
1 4 9 7 3 8 7 2 8 7 3 0 0 6 5 3
0 6 1 7 1 3 7 5 0 1 5 1 6 9 5 5
0 0 3 8 0 4 0 6 0 0 1 3 2 1 9 3
6 5 2 7 5 5 2 4 8 8 5 4 8 8 4 1
5 5 8 9 8 7 1 0 3 9 3 3 4 5 0 6
0 5 7 2 6 3 1 5 6 1 8 9 3 5 7 2
3 5 7 6 6 4 8 8 1 2 4 8 2 9 4 1
7 6 3 2 5 0 5 0 7 2 6 6 6 4 7 8
0 5 0 9 9 8 9 5 0 7 7 6 2 2 7 8
9 6 7 8 5 3 3 7 5 1 1 9 0 8 2 9
5 9 1 4 5 6 4 6 0 4 7 2 8 0 9 4
1 9 3 9 0 6 5 0 1 4 2 5 8 0 6 0
1 1 1 5 5 5 5 7 0 6 3 1 1 4 9 2
```

- ☐ 277470
- ☐ 770358
- ☐ 083617
- ☐ 510065
- ☐ 159073
- ☐ 531579
- ☐ 211859
- ☐ 439275
- ☐ 123991
- ☐ 364310
- ☐ 304314
- ☐ 914564
- ☐ 881741
- ☐ 920053
- ☐ 710628
- ☐ 505635

LARGE PRINT
NUMBER SEARCH
PUZZLES BOOK FOR ADULTS

VOLUME 1

SOLUTIONS

◆ ◆ ◆

Dear Valued Puzzler,

Thank you for purchasing our number-search puzzle book!

If you enjoyed it, we would greatly appreciate it if you could leave a review on Amazon. Your feedback is invaluable in helping others and allowing us to continue creating high-quality products.

Thank you for your support!

PUZZLED 🄟 PARROT

Puzzle 1

```
1 1 1 4 7 7 0 6 3 2 4 7 1 6 3 7
9 2 2 0 9 7 8 8 6 4 7 8 8 2 1 5
2 3 7 3 4 4 5 2 2 0 4 8 6 1 8 2
6 9 5 4 7 6 7 7 0 2 2 5 2 5 6 7
9 9 6 1 6 5 2 6 5 5 2 6 0 8 2 3
6 4 1 9 6 2 9 1 4 6 0 0 0 4 0 4
8 2 5 8 6 7 5 0 0 8 6 7 6 3 9 7
8 1 9 3 3 3 5 9 9 7 3 2 6 4 0 8
8 5 5 7 8 9 7 4 0 6 9 6 6 6 5 1
4 0 9 6 7 8 7 9 7 8 6 1 2 5 6 9
5 5 6 4 3 7 7 8 2 1 5 7 7 8 4 4
2 3 7 8 2 1 6 4 7 3 5 8 4 4 2 8
6 5 4 8 3 2 7 1 6 9 8 1 4 8 2 2
9 6 8 3 0 2 2 7 3 1 1 1 0 5 5 9
1 5 5 2 2 0 7 9 0 5 6 1 8 4 5 3
9 1 2 7 2 3 0 1 6 9 5 5 3 4 9 5
7 3 5 0 2 9 2 3 8 4 1 7 3 2 7 8
1 5 6 7 0 4 9 3 6 1 5 2 4 3 5 4
```

- [ ] 792381
- [ ] 167547
- [ ] 967485
- [ ] 002681
- [ ] 742206
- [ ] 926968
- [ ] 384173
- [ ] 207767
- [ ] 051249
- [ ] 096674
- [ ] 846738
- [ ] 756159
- [ ] 032721
- [ ] 962548
- [ ] 688790
- [ ] 197148

Puzzle 2

```
2 1 7 0 7 1 1 0 5 1 8 0 3 2 7 2
4 9 8 3 8 4 5 5 3 1 8 1 2 7 0 3
3 7 6 2 5 0 9 7 5 4 7 0 7 7 0 2
5 2 1 0 8 1 1 2 4 3 3 3 9 8 0 3
4 6 4 6 1 3 0 8 5 6 8 0 9 2 7 9
4 1 0 6 8 9 4 3 5 1 0 6 3 8 6 0
8 4 4 8 3 5 3 7 2 0 4 1 7 7 4 7
1 0 4 3 2 2 4 2 4 4 9 3 3 5 6 6
3 7 3 6 0 1 2 3 3 4 9 7 1 2 3 6
7 5 4 6 8 4 6 4 8 9 6 4 9 8 3 6
6 7 2 5 7 3 8 6 6 7 8 6 2 8 2 2
6 6 5 1 5 4 6 1 2 3 5 2 9 3 3 5
9 8 2 7 8 3 0 8 0 4 1 4 9 5 0 1
6 8 7 8 1 7 8 4 0 3 9 8 6 7 8 1
3 2 4 2 5 9 2 6 3 8 4 0 1 7 8 6
0 8 3 6 6 3 3 4 1 7 3 2 0 0 9 0
8 6 9 2 8 5 5 6 6 1 6 9 6 9 7 6
4 9 5 1 6 5 2 1 8 4 0 6 7 4 1 6
```

- [ ] 672573
- [ ] 850979
- [ ] 266128
- [ ] 355483
- [ ] 929137
- [ ] 178343
- [ ] 000764
- [ ] 041687
- [ ] 324259
- [ ] 683554
- [ ] 143048
- [ ] 084264
- [ ] 487187
- [ ] 474382
- [ ] 423015
- [ ] 633230

Puzzle 3

```
7 5 7 2 6 5 5 3 2 0 9 0 4 0 8 9
8 7 9 8 1 3 5 5 1 5 1 8 3 0 1 9
3 6 9 5 5 3 0 5 0 1 1 9 6 2 3 6
5 8 4 7 3 6 1 7 9 5 1 4 8 8 9 6
3 6 5 3 8 9 6 4 4 8 5 7 1 7 4 7
7 4 7 9 3 1 7 1 1 6 9 6 4 3 8 7
9 3 5 9 8 0 9 0 2 3 1 9 8 9 1 4
4 5 8 5 4 1 7 2 3 3 9 8 2 3 0 4
0 9 9 7 0 4 0 5 5 5 1 6 6 3 0 6
6 8 7 7 5 9 4 6 7 7 2 9 4 5 3 2
5 4 3 1 7 0 2 4 5 3 5 6 5 0 7
9 1 7 9 8 3 1 0 5 9 0 6 6 3 0 1
0 6 3 3 6 7 0 1 0 4 2 0 6 7 5 2
6 7 3 2 6 6 7 9 7 1 6 9 5 0 6 1
4 9 1 8 5 1 2 2 6 2 8 2 6 1 2 4
8 1 2 7 0 7 4 4 4 8 8 2 9 7 3 8
5 6 0 1 1 5 9 4 7 4 5 5 7 1 4 2
2 6 1 9 2 8 7 8 0 5 3 7 0 6 9 9
```

- [ ] 163748
- [ ] 649577
- [ ] 757500
- [ ] 560497
- [ ] 745571
- [ ] 989923
- [ ] 769294
- [ ] 809023
- [ ] 389644
- [ ] 983105
- [ ] 878910
- [ ] 468675
- [ ] 369553
- [ ] 056816
- [ ] 444882
- [ ] 951106

Puzzle 4

```
3 7 9 8 7 6 2 6 7 8 7 2 5 9 9 7
9 7 2 7 5 8 5 9 7 5 8 3 2 6 9 0
5 3 2 2 8 0 7 4 0 6 7 5 9 6 5 9
0 6 6 3 3 7 0 3 6 0 2 6 2 5 9 5
5 4 7 4 2 8 1 9 8 6 4 4 3 0 2 1
5 9 6 7 4 6 8 8 6 8 1 5 9 6 4 5
3 6 6 2 5 7 3 7 8 1 1 3 6 1 7 0
7 8 9 6 9 8 5 4 8 7 2 4 9 1 0 0
2 4 6 1 2 9 9 6 6 5 7 2 4 3 7 7
5 6 4 6 9 8 1 8 2 5 8 8 3 7 9 4
7 4 7 1 7 6 7 1 8 7 9 0 8 5 6 7
2 8 4 4 2 3 1 9 9 0 1 3 7 3 2 3
6 4 2 3 3 7 9 8 7 3 3 8 1 9 3 5
1 2 1 6 2 2 0 9 9 8 1 3 1 5 7 6
2 2 3 1 3 0 2 6 4 8 2 2 9 5 6 4
2 6 0 3 4 6 6 0 6 5 8 9 1 7 0 1
7 6 3 1 0 9 9 4 3 7 7 5 0 7 1 7
1 8 2 1 4 8 1 4 5 7 0 8 0 9 8 1
```

- [ ] 677589
- [ ] 412706
- [ ] 956436
- [ ] 546532
- [ ] 364475
- [ ] 876267
- [ ] 395055
- [ ] 981315
- [ ] 622484
- [ ] 376683
- [ ] 281986
- [ ] 308243
- [ ] 280740
- [ ] 627527
- [ ] 429599
- [ ] 734990

Puzzle 5

```
0 4 6 7 6 3 3 7 7 3 4 6 2 3 0 6
4 5 9 1 7 7 3 3 4 4 6 2 4 5 7 8
7 7 4 0 1 1 3 6 1 5 0 1 1 7 6 9
4 2 2 2 2 9 8 9 6 6 0 1 4 1 7 5
8 2 9 7 7 1 4 0 3 0 1 0 8 4 9 5
4 2 0 1 1 8 4 5 9 9 5 0 8 8 6 0
0 0 2 5 7 3 5 4 6 0 2 1 8 1 9 1
8 5 5 2 8 5 9 9 9 4 0 0 2 3 6 6
3 9 7 4 2 8 5 3 7 2 6 9 7 4 4 6
7 4 3 2 1 6 9 0 8 1 5 4 3 6 0 0
2 8 9 0 5 4 3 4 5 5 9 7 6 8 5 8
8 7 3 3 1 7 4 8 9 0 0 8 2 0 1 9
5 4 1 5 8 6 3 5 1 8 0 7 6 1 8 7
4 1 0 0 2 6 1 4 5 5 6 3 6 4 4 8
3 8 7 4 5 0 9 5 0 9 5 0 4 1 3 0
5 7 2 7 5 9 3 7 4 0 6 9 5 7 6 3
4 8 2 1 6 9 0 5 9 3 4 0 0 6 6 2
7 5 7 3 3 6 4 7 8 8 1 4 3 9 3 0
```

- [ ] 771809
- [ ] 098471
- [ ] 806464
- [ ] 735460
- [ ] 345827
- [ ] 095095
- [ ] 251006
- [ ] 375275
- [ ] 697670
- [ ] 205889
- [ ] 385076
- [ ] 884142
- [ ] 673939
- [ ] 542785
- [ ] 689179
- [ ] 593400

Puzzle 6

```
9 2 0 0 9 1 1 1 5 5 1 3 1 6 0 6
7 3 0 1 8 6 7 1 1 9 6 0 6 3 4 7
2 4 8 7 5 9 9 9 2 9 0 4 6 7 1 1
2 1 1 1 0 2 9 6 4 3 7 1 0 2 4 4
1 2 7 8 0 4 6 2 4 6 7 5 6 1 1 9
9 9 4 9 2 5 2 4 1 9 6 7 7 6 8 4
0 2 5 2 1 1 3 5 3 5 6 7 6 8 5 7
1 7 7 7 7 9 9 9 8 9 7 5 1 7 3 4
5 8 0 0 9 5 2 3 1 3 0 0 3 3 9 4
2 6 0 6 0 4 7 3 3 7 2 2 3 4 9 5
3 8 1 5 5 5 5 0 2 6 9 3 1 8 4 5
6 4 6 6 5 8 3 3 8 1 0 4 4 2 0 4
6 5 8 2 7 2 3 4 6 9 0 1 5 6 5 8
6 0 3 0 8 8 2 2 7 4 9 2 5 3 9 4
1 3 7 5 2 2 3 8 9 3 2 9 3 9 6 5
5 7 2 2 3 1 1 6 7 7 0 2 9 6 6 2
0 6 5 1 8 9 9 4 4 7 1 3 4 4 4 8
5 4 0 9 8 4 3 7 4 2 9 5 3 0 2 9
```

- [ ] 232919
- [ ] 231167
- [ ] 427570
- [ ] 957945
- [ ] 091115
- [ ] 799623
- [ ] 061315
- [ ] 077667
- [ ] 395963
- [ ] 049935
- [ ] 673054
- [ ] 822550
- [ ] 722190
- [ ] 029009
- [ ] 603088
- [ ] 660676

Puzzle 7

```
3 0 1 2 1 7 7 0 8 0 3 1 9 3 8 3
1 1 4 4 6 0 0 6 9 6 5 3 5 1 3 8
0 9 1 9 7 5 8 0 0 3 2 3 4 5 9 4
2 2 6 9 6 0 1 9 3 7 4 0 9 0 9 3
7 6 3 3 9 4 8 5 3 7 3 4 5 7 4 4
7 8 0 8 5 4 4 9 7 7 5 1 8 2 2 9
8 8 5 2 7 3 0 6 7 3 9 2 3 8 8 0
8 2 1 8 2 0 1 6 6 0 8 8 6 4 1 1
7 2 7 8 1 7 9 5 1 8 6 4 2 7 8 5
1 4 5 7 8 2 0 8 0 8 8 8 2 8 9 5
4 7 2 2 9 6 7 8 7 1 8 6 3 6 1 3
6 9 8 4 2 8 4 1 6 0 6 5 8 4 7 5
0 1 8 2 4 1 1 1 8 7 4 1 1 7 3 0
9 9 0 0 3 4 8 7 1 4 8 7 5 7 3 6
3 0 4 0 4 1 8 1 0 4 9 0 8 7 7 0
5 0 8 4 2 9 2 5 7 8 3 6 7 0 0 3
1 1 7 2 7 8 2 7 1 1 5 6 0 7 1 5
4 0 2 2 8 9 7 1 7 6 8 4 5 9 7 7
```

- [ ] 437358
- [ ] 802875
- [ ] 468806
- [ ] 984284
- [ ] 832937
- [ ] 181404
- [ ] 150728
- [ ] 157794
- [ ] 940618
- [ ] 163681
- [ ] 872468
- [ ] 839130
- [ ] 196353
- [ ] 887720
- [ ] 102812
- [ ] 783670

Puzzle 8

```
9 6 9 9 6 3 8 1 0 4 8 4 3 2 2 9
1 5 2 8 3 3 5 0 5 7 5 4 8 1 4 2
2 8 3 1 6 3 8 0 2 8 7 0 4 6 9 5
1 1 4 2 8 9 1 8 8 5 4 7 8 5 6 6
9 2 4 9 7 2 7 7 0 9 7 4 0 1 2 7
9 9 4 2 0 8 4 5 1 5 9 5 5 9 1 6
7 1 0 0 4 7 1 4 8 7 7 6 0 4 1 7
2 9 0 2 5 4 4 3 6 8 8 3 5 7 4 4
6 0 0 4 5 0 6 4 9 3 1 4 9 2 8 1
4 6 2 9 8 6 3 5 4 8 9 2 9 6 1 2
0 1 7 6 9 7 4 0 4 2 4 4 0 2 3 6
0 7 8 5 7 1 9 6 5 5 1 7 2 6 7 3
8 2 7 5 8 0 0 8 1 2 6 3 1 8 0 1
2 0 3 5 3 6 5 6 7 7 8 9 2 9 2 0
5 5 7 4 4 2 8 0 4 9 2 4 9 6 0 1
9 1 3 2 3 9 0 4 0 4 6 7 4 3 1 2
1 5 7 4 3 7 2 7 2 0 3 4 9 7 6 0
3 2 3 0 8 8 0 3 3 1 8 3 3 4 9 4
```

- [ ] 008126
- [ ] 442804
- [ ] 364147
- [ ] 749156
- [ ] 004627
- [ ] 440745
- [ ] 040467
- [ ] 292189
- [ ] 635302
- [ ] 824936
- [ ] 496211
- [ ] 025646
- [ ] 457800
- [ ] 574797
- [ ] 784019
- [ ] 050421

Puzzle 9

```
6 9 2 7 3 0 2 2 8 2 8 6 4 6 8 2
0 1 9 4 0 0 3 4 8 6 1 7 2 3 9 3
4 0 0 1 7 9 1 6 2 6 6 2 4 1 2 7
5 1 4 4 6 8 4 4 7 5 0 3 8 2 6 0
2 6 5 8 7 6 6 5 2 4 6 0 5 0 2 5
7 6 4 7 9 8 2 2 1 6 5 5 8 8 0 7
9 2 4 4 3 2 5 3 2 1 2 8 3 6 4 6
4 0 8 4 2 5 0 8 4 1 2 4 8 3 4 6
1 5 3 0 1 1 9 4 8 4 6 4 0 1 3 0
4 3 3 9 1 4 7 1 7 4 1 5 5 6 9 9
1 8 3 9 9 3 1 2 6 0 5 4 2 6 4 1
0 9 2 3 7 3 6 9 1 8 7 6 1 0 4 5
7 2 9 1 7 5 0 4 8 5 3 2 7 8 0 2
1 9 8 1 0 5 0 1 1 2 1 8 1 3 9 9
0 0 5 3 8 3 9 3 6 0 4 6 8 2 4 2
0 6 6 5 6 3 3 2 3 6 3 6 3 9 1 6
6 3 6 0 1 4 5 6 7 9 7 9 8 3 7 1
9 7 9 9 4 6 8 1 8 7 7 1 3 7 0 0
```

- [ ] 662053
- [ ] 866358
- [ ] 489204
- [ ] 432661
- [ ] 844750
- [ ] 376548
- [ ] 062139
- [ ] 174155
- [ ] 791626
- [ ] 301284
- [ ] 930146
- [ ] 256122
- [ ] 858740
- [ ] 014567
- [ ] 463821
- [ ] 305793

Puzzle 10

```
7 4 2 2 7 5 2 1 7 3 0 1 8 5 3 2
0 6 4 6 1 9 7 5 6 5 8 9 3 3 6 1
5 6 3 4 1 9 9 9 1 4 7 4 1 6 5 0
2 2 1 7 8 3 6 1 6 4 8 9 8 2 4 8
1 7 0 4 5 9 0 6 6 4 4 8 1 4 6 0
3 2 8 7 8 4 3 0 7 8 1 1 0 0 6 1
8 2 0 2 3 5 7 2 6 0 9 7 4 7 3 3
5 6 5 2 3 9 7 4 1 0 8 2 0 7 3 4
3 4 0 5 6 6 8 6 5 7 1 5 5 1 8 5
5 4 3 7 0 4 4 4 3 7 8 8 2 5 0 2
9 5 6 3 0 3 9 0 8 7 1 8 2 7 1 3
6 6 6 4 8 9 7 9 8 3 3 0 8 3 1 8
9 7 1 2 6 0 0 7 6 3 1 8 5 5 7 6
0 6 8 6 7 9 0 3 4 8 9 7 8 3 8 9
6 3 0 1 0 3 5 9 5 8 0 3 1 7 8 8
8 7 4 5 7 9 1 7 2 0 7 8 3 3 0 4
3 3 9 0 1 8 2 9 0 0 9 0 1 3 4 5
3 6 7 0 0 7 9 7 4 0 4 0 4 1 6 0
```

- [ ] 543704
- [ ] 571053
- [ ] 281060
- [ ] 948773
- [ ] 759646
- [ ] 466095
- [ ] 173018
- [ ] 339018
- [ ] 008833
- [ ] 833664
- [ ] 361648
- [ ] 029008
- [ ] 985657
- [ ] 630103
- [ ] 563419
- [ ] 001187

Puzzle 11

```
9 4 3 4 0 7 8 5 5 1 9 2 6 3 7 9
1 6 4 5 4 7 4 6 6 1 7 3 2 2 8 3
8 9 6 4 4 7 3 0 3 8 0 4 0 6 6 7
7 1 4 7 7 1 5 1 3 4 0 5 5 7 4 0
8 6 0 7 4 4 6 0 8 8 8 4 8 8 9 9
9 5 3 4 5 0 4 3 2 0 4 9 5 7 5 7
7 5 6 0 3 6 1 6 9 2 5 9 5 0 6 7
8 5 6 9 3 9 9 2 8 3 1 0 0 2 4 1
5 3 0 9 1 2 3 6 8 3 2 1 8 2 3 5
4 1 9 2 9 8 0 6 4 8 4 5 5 7 9 4
7 1 0 8 9 0 0 6 9 0 1 2 6 8 6 6
5 3 0 1 3 0 8 7 9 3 5 5 8 9 2 9
5 5 9 0 3 9 4 5 4 4 5 5 5 2 8 2
2 7 9 4 8 7 6 7 4 4 9 0 8 1 4 2
7 4 7 6 0 3 4 5 1 1 5 0 0 4 9 5
3 2 5 0 2 4 2 7 0 7 5 0 4 6 5 0
5 7 6 9 9 0 1 7 2 0 8 4 4 7 2 6
7 1 2 7 1 2 9 0 1 1 0 9 1 3 2 7
```

- [ ] 563382
- [ ] 632190
- [ ] 904774
- [ ] 346403
- [ ] 099675
- [ ] 361692
- [ ] 155188
- [ ] 400511
- [ ] 710890
- [ ] 785475
- [ ] 242052
- [ ] 962849
- [ ] 513405
- [ ] 988488
- [ ] 192980
- [ ] 590394

Puzzle 12

```
7 4 3 9 4 5 2 6 0 8 7 7 0 4 8 6
1 9 5 4 4 5 3 0 8 6 5 0 1 5 0 0
2 0 4 5 1 6 2 9 6 8 8 5 9 3 4 7
8 2 7 0 5 2 1 3 3 7 3 1 4 4 9 3
4 9 5 3 3 2 8 7 7 7 4 0 1 4 6 8
6 5 5 4 1 5 6 3 9 6 3 2 8 8 7 6
2 5 0 2 5 6 7 8 4 4 7 9 3 0 4 8
9 3 3 7 8 3 7 3 6 8 9 8 5 5 7 9
4 2 5 2 6 9 8 1 4 9 5 3 5 1 4 4
3 6 6 4 1 8 1 9 0 2 8 3 4 2 5 5
3 3 8 1 4 1 4 7 3 3 2 9 4 4 4 6
0 3 8 2 0 0 6 3 6 4 0 6 5 9 2 6
0 0 2 0 6 6 5 6 0 4 8 3 6 2 4 7
4 9 3 3 2 6 2 6 9 8 6 2 2 7 9 0
3 8 3 4 6 6 2 1 2 5 6 4 6 0 4 3
8 5 6 2 0 9 6 5 2 9 4 7 5 0 2 0
6 7 5 8 0 0 1 6 2 1 9 0 6 8 0 3
3 1 2 5 5 3 9 9 4 8 4 7 1 8 0 5
```

- [ ] 554156
- [ ] 688308
- [ ] 212601
- [ ] 953328
- [ ] 046497
- [ ] 886230
- [ ] 854646
- [ ] 355445
- [ ] 124927
- [ ] 312507
- [ ] 937393
- [ ] 843086
- [ ] 967474
- [ ] 596611
- [ ] 508443
- [ ] 834003

Puzzle 13

```
9 2 6 0 9 1 4 1 5 0 0 0 4 7 7 5
4 9 3 6 8 4 7 1 2 8 1 6 3 3 7 3
4 3 0 0 5 6 3 0 0 3 8 5 3 1 4 3
7 1 6 9 5 3 7 6 3 1 9 4 6 7 1 4
7 4 4 9 0 5 4 3 9 5 7 4 2 3 5 6
4 5 4 2 5 9 6 0 3 4 1 4 1 8 9 6
3 7 6 3 4 5 6 1 4 3 8 8 2 1 0 2
7 5 5 6 7 6 6 8 3 7 6 3 9 2 5 0
3 0 8 1 8 7 6 9 3 1 9 8 9 4 7 0
4 2 8 3 0 6 9 2 6 8 2 8 8 1 2 0
6 9 0 3 7 2 6 0 6 0 8 0 8 5 7 1
2 5 5 9 9 4 2 3 7 9 1 4 0 1 5 6
4 6 7 7 3 6 4 8 7 5 6 4 1 5 3 1
6 3 0 4 4 2 8 7 5 0 0 3 5 6 9 8
1 4 7 1 0 1 3 2 4 2 4 4 7 3 8 8
2 8 4 1 9 4 1 9 2 7 1 6 6 3 8 3
6 0 5 5 9 3 6 1 7 6 2 7 7 9 5 0
9 9 2 0 7 6 3 4 7 3 4 1 8 9 8 1
```

- [ ] 749891
- [ ] 613842
- [ ] 724577
- [ ] 114793
- [ ] 359567
- [ ] 301892
- [ ] 384963
- [ ] 609923
- [ ] 668663
- [ ] 608085
- [ ] 703518
- [ ] 415000
- [ ] 188286
- [ ] 114759
- [ ] 314575
- [ ] 413583

Puzzle 14

```
2 5 1 5 7 1 0 4 6 2 1 5 3 7 0 5
1 9 5 8 0 7 0 5 2 7 3 7 5 4 6 4
8 5 5 3 8 6 2 0 1 1 2 3 8 1 2 7
4 1 8 8 5 5 9 0 8 2 8 1 9 0 6 7
3 7 3 6 1 8 1 5 8 2 5 6 2 7 4 1
0 4 4 7 5 3 4 1 9 1 5 7 7 3 7 8
2 2 0 8 0 0 5 6 0 5 7 4 1 6 3 2
6 1 3 4 3 4 2 2 2 1 0 2 8 5 6 9
4 6 1 1 1 2 2 8 0 0 9 7 2 2 9 5
3 3 4 4 4 8 0 3 8 1 7 1 6 1 9 8
2 4 0 7 8 9 6 1 8 7 8 3 2 7 3
5 6 4 0 6 6 7 5 9 2 2 2 8 6 4 1
1 5 4 8 0 7 0 6 9 5 5 6 2 7 3 7
5 2 0 4 4 9 2 6 1 0 4 5 1 1 2 5
1 6 6 1 4 2 6 0 0 7 2 1 8 6 5 8
2 8 3 1 7 9 7 7 9 3 5 6 0 7 1 7
7 8 8 6 3 1 9 3 7 2 0 9 9 5 8 9
3 5 8 5 4 7 4 7 1 4 9 9 0 0 0 4
```

- [ ] 851101
- [ ] 245287
- [ ] 900034
- [ ] 855709
- [ ] 276746
- [ ] 974325
- [ ] 122243
- [ ] 362817
- [ ] 836044
- [ ] 149761
- [ ] 729853
- [ ] 742713
- [ ] 684112
- [ ] 637462
- [ ] 365212
- [ ] 070527

Puzzle 15

```
5 6 3 5 5 8 9 4 0 2 8 2 1 0 2 7
4 4 9 7 6 0 1 7 2 7 0 3 7 2 7 2
5 2 5 5 9 7 5 9 9 6 8 2 1 2 5 8
7 8 6 2 8 1 1 4 5 0 8 9 0 3 4 2
4 4 8 4 4 8 5 0 6 8 0 8 6 6 4 9
1 4 3 4 8 2 1 7 3 8 1 3 0 0 2 0
7 3 9 5 8 5 1 0 2 0 6 9 3 5 1 6
8 5 8 0 6 3 3 3 3 3 2 0 4 6 8 6
7 8 0 6 2 9 2 5 7 1 5 2 6 9 7 6
9 0 7 6 6 8 7 3 2 2 9 2 7 5 1 7
1 1 7 5 4 1 1 3 8 7 2 3 7 4 9 9
1 2 9 5 7 0 3 0 6 9 0 9 4 8 8 0
6 6 4 2 9 1 0 7 8 6 9 8 0 9 2 2
0 3 9 9 3 8 7 4 8 4 4 8 0 3 5 4
3 8 5 5 5 5 2 3 5 4 9 0 0 0 3 2
9 9 5 2 2 0 5 4 7 0 7 8 0 4 7 6
1 6 8 3 6 9 1 1 4 5 0 9 8 1 5 3
9 2 9 2 3 0 8 8 1 3 4 0 6 5 6 0
```

- [ ] 995220
- [ ] 633009
- [ ] 808893
- [ ] 096030
- [ ] 563237
- [ ] 974626
- [ ] 306119
- [ ] 022360
- [ ] 545242
- [ ] 307271
- [ ] 801263
- [ ] 871982
- [ ] 466379
- [ ] 635589
- [ ] 445066
- [ ] 555523

Puzzle 16

```
2 5 4 1 2 8 8 6 2 4 9 7 4 0 6 5
8 6 0 9 9 6 0 3 4 2 2 5 8 1 7 9
5 0 3 9 7 4 8 7 2 5 1 6 1 0 2 3
9 7 7 8 7 0 5 2 7 9 8 0 3 6 4 6
5 3 0 5 3 1 3 9 2 0 2 1 5 5 4 4
5 6 6 5 6 1 8 6 1 9 2 6 7 5 2 8
4 2 3 6 2 5 9 9 4 2 3 6 1 9 6 0
0 8 2 6 4 2 7 2 8 8 0 8 5 7 8 7
8 3 3 7 5 9 7 1 5 3 0 5 5 3 4 2
6 2 4 7 0 6 6 8 4 4 1 5 3 3 0 2
4 8 0 9 2 3 4 4 4 7 8 7 8 4 9 1
1 7 5 3 8 5 1 6 8 0 1 5 5 0 0 4
3 8 2 7 7 7 4 0 5 3 7 5 4 2 7 1
8 9 4 3 7 1 3 5 2 4 9 0 9 7 9 8
4 4 6 5 9 2 2 4 2 3 8 0 6 7 6 8
4 5 1 2 4 6 7 0 6 1 5 1 0 4 3 9
3 7 6 7 8 1 4 3 3 4 4 8 0 7 2 3
4 5 6 8 9 3 3 2 0 4 0 1 7 3 0 6
```

- [ ] 548547
- [ ] 127242
- [ ] 403706
- [ ] 472341
- [ ] 595540
- [ ] 423619
- [ ] 704872
- [ ] 090796
- [ ] 566779
- [ ] 655012
- [ ] 238263
- [ ] 963571
- [ ] 372969
- [ ] 288085
- [ ] 234052
- [ ] 736245

Puzzle 17

```
6 9 6 4 0 2 6 2 5 8 3 7 9 5 4 5
4 6 1 9 8 9 1 7 7 9 5 3 7 3 7 7
7 4 8 0 3 8 9 8 0 9 3 5 5 8 1 7
2 1 1 0 2 7 3 2 7 2 4 9 7 3 1 0
1 5 3 0 3 8 1 5 7 6 7 4 2 3 1 9
7 5 0 3 6 4 8 0 9 4 2 2 2 4 0 6
5 8 1 3 8 9 3 9 2 5 9 7 2 6 9 9
4 7 6 3 6 5 3 2 6 6 3 2 8 7 0 4
4 5 9 4 7 7 7 9 2 9 3 7 8 6 6 6
4 8 2 2 5 0 9 4 8 2 5 0 5 8 0 5
4 4 6 0 6 4 6 2 5 8 5 2 8 7 0 9
0 7 6 8 2 1 6 5 2 0 9 8 4 6 0 6
1 3 1 3 6 7 5 7 4 6 4 8 5 5 7 9
6 8 6 3 0 3 2 8 1 1 0 7 6 0 1 9
4 9 4 7 4 5 7 1 6 9 4 0 0 0 2 7
4 3 9 6 4 1 4 4 5 4 7 6 6 2 4 0
9 9 9 1 5 9 6 3 3 2 7 1 1 6 3 9
2 6 8 7 2 5 8 0 4 8 9 8 2 9 7 2
```

- [ ] 992749
- [ ] 867562
- [ ] 490003
- [ ] 752492
- [ ] 083236
- [ ] 272076
- [ ] 714075
- [ ] 926285
- [ ] 000496
- [ ] 217000
- [ ] 643383
- [ ] 947457
- [ ] 368684
- [ ] 266745
- [ ] 788207
- [ ] 975722

Puzzle 18

```
3 2 7 3 9 4 1 9 5 3 7 9 0 0 6 8
7 7 8 4 2 3 1 9 8 9 9 6 1 6 4 6
2 7 7 8 7 8 6 6 1 1 0 8 9 3 4 2
9 5 0 8 5 6 7 5 9 6 3 8 7 4 0 7
9 6 0 0 4 2 2 1 2 0 8 1 7 1 0 9
6 5 8 9 8 0 1 4 8 2 9 9 3 9 5 7
6 9 8 5 8 8 5 0 9 3 0 7 1 9 7 7
0 0 1 1 2 1 1 0 4 9 7 4 1 8 6 0
1 4 9 5 8 0 9 7 3 3 5 1 9 4 4 0
4 3 5 6 6 3 1 1 6 6 7 0 6 2 7 8
1 0 4 1 5 9 3 9 3 1 7 3 2 9 9 1
8 5 5 4 4 0 6 2 6 4 9 8 4 5 9 3
7 8 6 4 9 2 1 6 0 8 8 1 5 6 9 1
2 5 8 2 3 2 2 2 1 8 6 1 4 1 9 5
4 1 6 8 7 1 5 3 6 8 4 3 5 5 0 6
0 3 9 8 8 1 1 2 0 1 3 1 7 3 1 4
0 7 3 6 5 7 5 8 6 6 4 4 9 0 5 2
3 7 7 4 0 4 6 1 4 3 6 5 2 4 8 5
```

- [ ] 711220
- [ ] 140071
- [ ] 916023
- [ ] 669927
- [ ] 966183
- [ ] 102118
- [ ] 073686
- [ ] 396865
- [ ] 142230
- [ ] 187240
- [ ] 215191
- [ ] 043058
- [ ] 335194
- [ ] 148023
- [ ] 778786
- [ ] 842319

Puzzle 19

```
7 8 3 5 9 6 8 4 0 5 9 6 1 0 5 0
7 0 1 3 0 9 6 3 9 8 7 8 3 7 2 7
4 3 7 6 4 8 0 2 0 5 5 5 6 5 3 2
6 7 7 5 0 7 0 8 1 9 7 3 2 5 6 9
0 5 8 7 1 5 2 7 3 7 4 7 7 4 6 7
7 4 0 4 2 9 5 0 8 8 6 1 0 6 1 9
1 4 5 3 1 3 0 8 0 2 7 9 6 4 3 0
7 1 0 4 3 3 7 7 2 9 4 9 9 9 6 0
7 8 9 9 0 8 3 5 8 7 7 7 1 7 5 2
5 9 0 1 6 3 4 3 1 0 9 4 6 3 0 8
5 9 1 5 2 6 5 6 3 7 6 4 8 2 5 8
3 0 5 1 8 5 6 9 7 8 3 3 6 1 2 3
3 3 0 9 8 7 5 3 7 1 8 7 2 7 4 2
3 6 9 6 9 3 6 5 6 8 8 1 3 2 9 8
6 1 2 4 0 8 6 3 1 9 5 1 4 6 9 7
8 4 0 8 4 8 9 7 8 1 3 4 3 5 0 4
6 2 0 6 2 7 4 5 7 8 0 4 6 1 0 2
7 5 7 3 3 6 6 0 1 1 5 3 9 3 6 9
```

- [ ] 907806
- [ ] 942505
- [ ] 405961
- [ ] 194347
- [ ] 507081
- [ ] 699949
- [ ] 754726
- [ ] 936903
- [ ] 240988
- [ ] 285506
- [ ] 415913
- [ ] 960726
- [ ] 537187
- [ ] 265637
- [ ] 387965
- [ ] 853719

Puzzle 20

```
1 3 1 3 7 7 7 5 0 0 3 6 7 2 8 3
1 8 8 6 9 1 6 8 4 2 9 0 2 4 5 1
3 5 6 5 9 6 2 1 9 5 8 6 1 6 8 6
6 3 1 0 3 1 9 6 5 8 7 7 2 0 1 7
2 7 5 4 6 1 7 0 1 2 3 1 4 9 3 1
2 2 9 0 6 9 8 0 6 5 1 9 7 5 7 5
8 3 3 8 6 4 7 0 6 1 9 2 9 1 1 7
7 5 0 3 7 1 4 9 0 1 2 8 0 0 7 9
7 9 7 7 1 8 2 6 1 7 2 4 8 2 4 7
1 9 1 4 9 1 8 6 3 4 4 1 8 3 7 3
4 5 3 3 2 2 3 0 3 0 0 0 1 0 8 3
4 0 8 5 5 1 3 2 3 6 6 7 7 2 3 4
7 5 2 2 7 8 8 9 8 4 2 9 8 7 7 5
9 9 2 4 0 9 4 5 2 0 9 0 4 6 3 1
3 4 6 3 3 5 3 0 7 6 6 9 5 0 0 9
1 5 3 1 4 8 9 0 5 3 8 5 1 6 6 6
7 0 0 7 3 7 7 9 7 1 9 0 2 1 4 2
3 8 7 6 8 7 0 5 5 2 1 9 3 3 6 8
```

- [ ] 896092
- [ ] 153148
- [ ] 603738
- [ ] 987880
- [ ] 569130
- [ ] 025490
- [ ] 005777
- [ ] 291378
- [ ] 516217
- [ ] 636205
- [ ] 797190
- [ ] 138226
- [ ] 397441
- [ ] 140786
- [ ] 581371
- [ ] 061211

Puzzle 21

```
0 3 6 7 1 6 9 4 6 1 8 4 7 6 5 8
6 5 7 3 9 0 5 1 6 4 0 0 8 2 9 3
6 7 0 3 3 1 0 5 1 8 5 4 6 0 8 7
8 0 9 6 2 2 5 7 9 2 8 1 7 9 5 0
6 6 3 5 6 7 2 7 4 4 5 9 5 6 0 5
4 7 4 7 0 6 2 5 2 0 3 3 0 5 6 4
5 1 9 8 5 9 8 6 1 5 6 7 8 2 7 8
7 1 0 0 8 3 4 3 0 2 4 8 5 7 9 9
2 9 5 6 1 4 8 1 2 6 8 4 5 4 6 7
3 7 0 4 0 8 0 8 4 1 8 3 1 4 3 8
1 7 1 1 9 5 1 4 8 9 2 6 1 0 9 1
1 7 6 8 5 1 4 0 4 0 6 0 8 4 0 9
3 0 9 9 4 7 4 4 6 2 8 6 8 9 7 6
3 3 7 6 3 2 5 8 1 5 8 9 3 4 1 3
9 2 7 1 0 4 5 1 5 3 7 6 0 6 7 8
6 9 1 1 9 6 6 9 7 8 0 4 4 9 9 7
8 4 7 3 5 3 4 2 3 9 1 4 6 2 1 7
3 1 1 7 0 1 9 0 4 3 4 0 8 4 1 3
```

- [ ] 144527
- [ ] 048136
- [ ] 986860
- [ ] 460875
- [ ] 573495
- [ ] 309947
- [ ] 945860
- [ ] 911253
- [ ] 657390
- [ ] 291840
- [ ] 914858
- [ ] 096527
- [ ] 950522
- [ ] 158934
- [ ] 468890
- [ ] 748164

Puzzle 22

```
8 9 2 5 0 2 5 4 8 2 8 8 5 8 9 4
3 6 7 8 5 6 8 6 0 4 2 0 4 9 2 6
8 4 2 3 0 6 5 5 6 0 4 9 6 2 9 8
7 0 7 5 5 1 3 0 1 9 0 7 5 6 5 1
0 2 9 9 5 1 5 9 9 5 3 5 6 9 1 1
6 0 0 9 0 1 0 7 0 4 4 1 4 4 4 0
2 8 7 2 7 7 2 4 0 4 2 3 6 3 1 4
7 1 0 1 5 4 8 6 1 1 7 0 9 7 7 1
5 4 9 6 9 7 9 4 0 2 7 9 1 8 5 6
1 8 7 7 9 9 4 8 6 8 8 7 5 3 8 1
1 4 3 3 0 9 8 3 4 9 6 0 4 5 8 8
3 4 4 3 8 4 3 2 6 0 1 3 6 2 5 8
4 7 9 4 8 5 1 3 8 9 8 9 7 8 8 1
4 2 8 6 3 6 1 0 7 2 0 8 2 4 4 8
6 1 1 0 3 5 6 5 9 1 5 7 8 4 3 4
5 9 7 5 7 0 1 5 2 5 0 9 7 4 0 6
5 8 5 6 9 0 3 6 1 4 6 0 2 4 9 0
1 2 2 8 1 6 1 2 9 1 1 2 1 3 0 1
```

- [ ] 140153
- [ ] 892694
- [ ] 513097
- [ ] 216733
- [ ] 070973
- [ ] 733880
- [ ] 208148
- [ ] 892502
- [ ] 295282
- [ ] 948880
- [ ] 069438
- [ ] 641630
- [ ] 367856
- [ ] 260863
- [ ] 799456
- [ ] 964656

Puzzle 23

```
9 9 7 8 7 5 5 0 5 5 3 6 2 9 4 7
0 5 8 8 7 1 8 6 6 7 1 6 5 0 0 9
9 4 3 8 5 9 8 3 1 1 7 7 9 3 3 7
6 6 9 7 6 8 9 9 2 6 5 7 4 3 1 0
0 4 2 2 1 9 7 8 4 1 7 2 5 0 4 9
7 5 8 5 0 2 2 0 2 4 0 3 4 3 6 5
9 7 0 3 9 7 2 4 5 3 1 4 0 6 5 5
5 2 7 7 5 1 0 5 9 5 8 6 5 1 9 1
4 8 5 5 4 6 6 1 4 6 7 0 9 2 1 0
5 0 0 8 0 0 8 4 7 3 2 6 6 7 0 1
4 2 2 6 2 9 7 4 3 2 6 5 5 9 4 6
0 6 1 4 3 6 4 9 5 4 9 8 2 2 6 1
1 2 4 5 8 4 7 7 1 3 7 2 2 0 4 9
2 6 2 2 1 0 7 9 0 8 9 3 5 2 1 8
8 0 9 9 1 6 3 3 2 8 3 7 0 9 4 6
8 4 6 9 5 3 7 0 1 8 8 1 2 1 5 7
6 7 2 0 5 0 5 2 7 2 2 0 5 2 8 5
1 7 1 2 8 2 3 3 4 9 1 5 3 5 6 6
```

- [ ] 684539
- [ ] 186671
- [ ] 875505
- [ ] 796278
- [ ] 706909
- [ ] 104545
- [ ] 210790
- [ ] 537018
- [ ] 343042
- [ ] 785870
- [ ] 465910
- [ ] 207017
- [ ] 978417
- [ ] 050527
- [ ] 122543
- [ ] 214296

Puzzle 24

```
8 1 6 1 7 7 6 7 7 4 6 9 9 8 7 9
7 0 5 7 8 2 7 3 0 2 3 0 1 7 8 9
2 4 7 5 3 2 7 4 0 9 9 0 7 0 2 0
9 7 9 3 2 0 5 2 7 0 9 9 2 7 9 4
7 2 4 7 9 7 4 8 3 4 3 5 8 5 4 8
3 2 8 7 2 9 9 5 0 7 4 6 0 6 8 8
9 1 8 4 7 9 4 5 6 6 5 4 7 0 4 9
9 3 0 9 2 1 1 4 5 6 8 6 6 8 3 8
0 9 2 2 4 8 1 9 6 5 7 4 3 7 2 9
4 1 9 0 8 3 0 9 4 6 0 4 9 0 3 1
1 6 9 1 1 5 8 7 2 9 9 8 5 0 7 6
5 0 7 7 0 1 9 9 1 7 7 6 0 5 0 2
9 3 9 1 1 2 6 8 4 8 8 0 3 9 2 8
5 7 9 3 2 9 6 9 1 9 7 7 7 0 6 4
1 5 3 4 0 3 9 1 6 7 4 3 9 0 8 8
7 3 7 8 1 0 1 1 0 1 0 2 8 5 2 1
7 2 0 8 2 2 2 5 7 3 1 4 7 6 4 6
4 8 7 3 2 8 4 2 4 5 7 8 2 4 5 4
```

- [ ] 870059
- [ ] 842290
- [ ] 787291
- [ ] 729194
- [ ] 055597
- [ ] 904766
- [ ] 287507
- [ ] 971991
- [ ] 159517
- [ ] 177472
- [ ] 689927
- [ ] 082225
- [ ] 116961
- [ ] 771618
- [ ] 843171
- [ ] 994466

Puzzle 25

```
6 0 0 2 5 5 9 7 0 8 1 7 2 3 3 3
6 1 1 4 9 5 0 9 1 1 1 7 4 3 7 4
9 4 5 5 7 9 5 7 6 7 8 2 4 2 8 4
5 6 4 4 5 9 5 7 3 8 4 1 6 4 1 3
1 5 3 3 2 9 8 0 4 5 4 3 8 5 8 5
0 0 0 6 2 1 3 2 0 1 5 1 3 1 2 7
7 8 4 5 7 8 9 7 0 6 3 6 7 7 2 7
4 5 1 8 5 2 0 2 8 8 9 3 9 2 7 8
3 2 8 8 0 0 3 8 4 8 2 1 6 8 4 5
4 6 4 1 9 5 8 5 9 3 0 8 7 4 8 2
3 8 9 6 4 1 3 9 5 1 9 6 7 8 0
4 1 5 3 6 5 4 0 9 8 8 9 4 5 4 0
6 9 9 9 0 5 1 7 9 1 2 8 9 3 3 9
4 6 6 4 5 4 1 4 4 5 5 0 4 0 6 5
8 5 3 6 3 6 0 9 9 6 1 6 5 8 3 0
8 0 7 6 1 2 5 4 7 7 1 5 1 3 0 4
3 9 0 2 0 7 0 8 2 1 3 3 3 4 9 1
4 9 6 3 4 2 0 1 2 9 2 1 0 3 0 9
```

- [ ] 055055
- [ ] 182511
- [ ] 309385
- [ ] 185388
- [ ] 018181
- [ ] 772131
- [ ] 958272
- [ ] 967649
- [ ] 644328
- [ ] 029354
- [ ] 791566
- [ ] 728475
- [ ] 403963
- [ ] 781822
- [ ] 319432
- [ ] 082133

Puzzle 26

```
9 2 1 1 2 2 8 8 3 6 3 1 9 5 1 3
6 0 2 1 0 4 4 1 3 7 5 4 7 3 0 5
6 5 1 6 8 3 2 4 1 4 9 3 2 5 3 9
8 3 9 0 4 1 0 3 0 5 8 6 5 6 3 4
2 7 5 0 8 5 5 5 5 4 5 6 7 1 6 5
7 4 1 3 8 5 8 0 2 8 2 6 4 3 4 0
5 1 3 5 4 1 6 7 2 8 5 8 1 5 4 0
8 9 3 4 7 6 1 2 0 0 4 5 5 7 4 2
6 0 3 1 2 4 9 8 6 0 9 4 9 8 6 3
9 0 8 1 1 3 8 4 5 9 8 1 4 9 8 4
0 1 3 4 4 8 0 0 0 5 9 0 8 7 9
8 4 8 1 9 6 7 5 1 9 1 4 8 4 5 9
0 4 6 6 0 3 9 0 4 6 5 1 2 7 5 9
2 2 6 0 2 6 9 1 1 3 3 4 8 5 6 1
2 9 5 5 8 6 2 2 3 8 6 0 1 7 7 2
7 9 5 2 6 7 8 4 8 2 5 2 1 0 5 7
1 9 6 2 6 7 5 3 6 6 9 2 9 6 9 7
5 3 5 1 4 4 5 7 7 4 7 1 4 9 3 2
```

- [ ] 748848
- [ ] 145904
- [ ] 221129
- [ ] 615513
- [ ] 039046
- [ ] 762597
- [ ] 149325
- [ ] 440428
- [ ] 940828
- [ ] 626753
- [ ] 235855
- [ ] 208096
- [ ] 598149
- [ ] 861988
- [ ] 260269
- [ ] 875567

Puzzle 27

```
0 6 8 8 1 2 1 5 2 3 4 1 5 6 0 3
0 4 9 8 7 3 2 7 0 7 2 4 3 2 2 5
1 3 3 1 0 6 8 0 0 0 3 1 1 4 3 7
3 8 7 0 2 1 7 9 2 3 4 9 9 7 7 1
3 4 8 9 6 3 5 0 4 4 6 6 3 5 0 0
7 3 9 1 1 4 8 4 9 3 8 1 3 8 4 6
4 9 4 6 4 8 6 2 0 4 2 1 7 5 9 7
8 8 5 8 5 7 2 6 6 7 2 5 1 3 4 2
7 2 0 9 7 4 8 8 2 2 3 7 8 5 4 3
5 9 5 9 6 4 2 2 3 8 8 3 8 3 2 4
1 1 3 2 7 7 9 6 8 5 5 2 4 1 8 5
2 2 0 5 7 5 7 7 4 8 4 0 3 8 6 7
6 7 9 2 9 6 6 5 1 7 7 3 1 3 8 3
9 1 2 0 8 1 6 6 8 6 3 6 2 9 6 0
4 4 2 4 1 5 4 5 0 2 9 7 9 9 2 1
1 6 0 5 1 8 6 8 6 6 4 4 9 2 4 2
9 9 0 1 2 0 6 9 2 2 0 0 6 2 1 4
0 9 7 8 9 0 2 2 4 3 9 1 7 5 1 9
```

- [ ] 255869
- [ ] 220248
- [ ] 704510
- [ ] 617947
- [ ] 496542
- [ ] 309220
- [ ] 382628
- [ ] 716307
- [ ] 430646
- [ ] 004635
- [ ] 490767
- [ ] 085165
- [ ] 664492
- [ ] 273938
- [ ] 049442
- [ ] 512694

Puzzle 28

```
6 3 8 8 9 6 8 7 1 0 9 2 6 9 7 7
1 3 1 5 7 4 5 4 6 1 8 1 6 9 9 6
7 1 7 7 9 6 9 0 8 7 6 7 7 3 0 0
2 9 3 6 6 6 8 5 1 2 6 5 5 9 6 2
3 8 3 9 5 4 8 2 2 1 7 7 3 0 4 3
0 1 4 8 3 3 0 0 3 3 6 1 7 2 0 9
2 5 0 1 6 3 6 4 0 2 4 5 7 1 8 6
7 6 6 9 4 8 7 6 4 9 0 7 8 2 8 6
9 1 9 0 4 4 7 9 5 2 5 5 0 8 7 0
5 5 7 7 6 9 2 1 2 8 7 0 8 0 0 3
2 8 1 7 5 6 2 4 7 8 3 7 7 9 1 4
5 2 6 0 4 7 4 2 6 0 1 1 5 3 4 8
4 6 4 4 5 4 6 4 1 6 0 2 4 5 8 6
3 7 8 0 6 8 3 8 5 1 4 2 5 6 0 4
6 5 6 8 9 9 1 6 5 8 7 1 5 0 5 3
5 7 6 6 3 8 9 7 7 2 1 4 0 1 0 5
8 2 7 7 3 8 3 9 0 2 9 4 7 5 1 8
8 8 7 6 1 4 8 8 9 3 0 7 6 4 1 7
```

- [ ] 463569
- [ ] 996468
- [ ] 768424
- [ ] 665831
- [ ] 390821
- [ ] 808754
- [ ] 786988
- [ ] 007178
- [ ] 967755
- [ ] 698190
- [ ] 721401
- [ ] 838865
- [ ] 505178
- [ ] 772440
- [ ] 547513
- [ ] 675377

Puzzle 29

```
1 7 9 0 8 8 9 9 9 7 4 4 8 1 6 7 7
5 0 4 8 9 0 4 2 5 1 3 7 6 7 4 0
4 5 4 8 5 8 0 5 7 9 3 5 4 0 7 2
3 1 4 3 6 5 1 6 6 4 6 9 0 3 5 5
0 7 0 1 7 6 0 9 7 6 6 9 6 3 6 0
1 3 7 0 0 0 7 7 7 1 4 2 6 5 1 2
6 3 6 3 1 4 2 7 5 3 5 7 8 5 6 1
6 4 1 4 5 9 6 6 6 5 8 3 7 1 9 5 0
5 3 7 3 9 8 0 7 4 7 3 3 7 5 8 1
3 0 2 5 6 2 2 7 4 1 1 9 9 3 2 9
5 4 7 9 8 3 2 4 7 6 1 2 8 7 9 7
9 1 6 1 7 9 5 8 1 8 1 2 7 0 3 5
3 9 7 2 3 4 2 7 3 1 2 6 5 1 0 9
7 3 4 4 4 1 2 3 9 7 2 9 0 8 6 6
6 5 4 8 8 8 6 5 8 3 9 7 7 5 3 9
9 0 4 9 4 2 6 6 5 9 1 2 4 1 3 7
3 2 7 6 8 8 8 8 5 2 0 4 0 3 2 1
1 0 8 1 1 1 3 8 5 5 0 5 7 9 2 5
```

☐ 701851	☐ 622741	☐ 704539	☐ 746654
☐ 484378	☐ 917385	☐ 767796	☐ 603928
☐ 641459	☐ 779385	☐ 543016	☐ 798324
☐ 962293	☐ 447998	☐ 040258	☐ 252266

Puzzle 30

```
8 7 4 6 6 6 0 0 7 4 7 2 1 1 2 5
3 3 7 5 1 3 6 7 8 0 6 9 6 7 8 8
6 6 3 2 5 2 6 8 3 4 3 6 7 6 7 4
4 7 5 2 9 4 2 2 6 8 4 4 7 9 2 6
5 6 5 4 6 9 5 4 3 4 9 5 8 9 4 4
4 7 8 2 2 4 8 1 5 7 6 0 7 8 3 2
4 9 6 9 3 5 7 8 1 7 1 2 5 7 8 3
1 6 8 1 4 9 5 7 0 4 3 3 3 4 2 0
1 9 8 4 7 0 8 8 4 7 2 6 8 2 2 3
3 4 2 6 8 4 6 9 2 0 2 3 6 7 2
5 8 8 8 5 2 1 9 0 6 6 3 6 4 9 5
2 4 7 7 4 7 7 6 4 9 1 0 9 8 5 8
1 0 2 2 0 4 4 1 9 4 5 1 1 1 2 9
2 7 0 0 9 0 5 2 0 9 5 8 5 2 5 3
2 1 3 0 6 8 0 3 0 1 6 2 6 3 1 3
8 0 6 1 7 6 7 2 9 3 4 2 2 1 9 9
3 4 0 0 7 4 3 1 0 2 2 9 0 3 3 5
1 9 9 7 3 9 8 1 1 7 1 6 7 4 0 9
```

☐ 070391	☐ 291468	☐ 224925	☐ 418786
☐ 243927	☐ 125888	☐ 764910	☐ 086031
☐ 862241	☐ 769987	☐ 450983	☐ 521228
☐ 962347	☐ 274700	☐ 367679	☐ 714857

Puzzle 31

```
5 6 7 4 4 1 9 1 1 1 5 7 6 4 5 3
3 2 9 1 9 4 2 5 8 3 6 9 5 5 6 7
3 5 4 2 3 9 3 1 2 5 6 2 7 6 7 1
5 8 5 0 5 3 3 6 4 4 8 7 7 5 6 9
3 2 1 7 6 6 0 8 7 9 8 6 5 7 3 0
6 2 7 6 5 2 7 7 9 0 4 3 9 1 4 9
8 9 7 7 2 5 7 9 1 5 1 8 7 6 1 7
3 4 1 4 2 0 8 4 6 7 1 8 3 3 0 4
9 2 3 1 4 6 7 6 7 4 1 2 1 9 4 0
6 0 5 3 5 3 1 7 6 3 0 4 3 9 8 5
0 6 5 6 3 4 9 6 6 5 6 1 9 5 5 4
9 4 6 4 1 4 9 3 1 4 2 5 6 5 1 0
0 4 3 8 8 0 4 8 9 4 6 9 8 0 7 6
1 0 0 2 7 8 8 8 3 8 8 5 3 6 1 9
9 9 7 7 0 4 3 8 7 0 5 7 4 3 9 7
7 6 8 9 5 4 6 4 6 0 7 1 2 3 3 7
5 0 5 0 2 1 3 0 3 4 1 8 7 7 6 5
5 8 5 5 2 9 8 4 3 6 7 5 1 4 5 9
```

☐ 772616	☐ 610886	☐ 402340	☐ 491923
☐ 494839	☐ 565241	☐ 964984	☐ 647866
☐ 764978	☐ 021303	☐ 191447	☐ 248373
☐ 524062	☐ 934036	☐ 104851	☐ 767265

Puzzle 32

```
9 4 5 6 7 1 3 2 3 7 0 4 9 9 4 9
8 2 7 9 2 5 7 3 1 3 5 2 5 1 4 8
8 9 4 9 0 0 6 5 2 6 0 8 0 6 6 2
4 0 5 9 4 8 9 5 0 7 2 7 8 4 0 8
2 9 1 9 4 6 5 3 1 1 8 2 3 0 0 3
0 5 3 6 8 8 3 8 1 1 4 1 2 1 2 9
3 0 6 4 9 6 5 5 8 1 4 0 1 2 2 5
0 4 1 4 2 3 0 8 3 2 8 4 0 0 0 1
1 3 1 4 7 0 4 3 1 4 1 7 3 3 1 1
3 7 2 8 1 5 3 0 0 4 4 3 1 4 5 5
4 6 5 7 0 1 7 2 8 9 1 8 1 9 2 0
8 2 8 5 2 9 1 9 4 3 7 9 7 8 7 4
9 6 8 1 1 7 6 1 1 3 6 5 6 0 9 5
5 5 7 8 2 7 2 1 1 8 7 2 1 7 3 1
5 8 9 2 7 0 9 8 3 7 0 6 0 1 5 8
9 2 5 8 1 2 2 0 4 2 8 2 3 8 0 0
9 1 3 6 3 1 4 3 5 0 6 7 1 9 3 5
6 1 6 1 7 1 1 8 2 6 4 6 3 1 3 6
```

☐ 804396	☐ 948950	☐ 204282	☐ 975746
☐ 711301	☐ 485814	☐ 916401	☐ 024889
☐ 525313	☐ 909504	☐ 353448	☐ 712781
☐ 072985	☐ 278824	☐ 631435	☐ 740127

Puzzle 33

```
8 7 1 0 7 4 6 3 6 8 5 2 7 3 6 0
0 5 6 0 1 7 6 0 2 0 4 1 8 4 3 1
8 0 3 5 1 1 1 4 1 6 7 1 8 9 2 8
5 5 4 0 6 4 3 2 7 2 3 1 4 3 5 4
5 4 1 1 7 6 8 3 0 2 0 9 3 4 8 5
3 7 4 7 1 4 5 1 3 0 6 7 0 5 6 8
6 4 5 0 9 9 4 8 4 7 8 4 8 0 1 8
6 5 7 9 7 1 6 4 2 4 5 8 5 8 5 5
2 0 4 9 6 8 2 6 4 0 4 0 2 2 7 4
0 8 1 1 0 6 0 2 0 2 6 7 6 3 1 3
3 4 4 0 4 6 4 4 4 9 8 4 9 1 3 8
5 8 9 9 2 1 3 8 6 3 2 4 1 7 8 2
1 3 5 3 3 1 2 9 8 2 1 5 6 6 6 5
5 4 4 4 2 5 0 0 7 5 4 7 6 8 7 9
9 1 0 4 1 7 0 5 2 8 4 8 2 8 7 0
2 5 7 9 8 2 8 8 2 8 5 8 5 6 6 1
7 9 7 4 7 7 3 1 4 2 4 1 6 2 2 2
3 4 3 2 7 1 6 5 3 9 0 9 3 3 2 0
```

- [ ] 417052
- [ ] 887021
- [ ] 676202
- [ ] 627464
- [ ] 882858
- [ ] 263900
- [ ] 601180
- [ ] 851791
- [ ] 457005
- [ ] 725863
- [ ] 440780
- [ ] 514384
- [ ] 773142
- [ ] 776831
- [ ] 035111
- [ ] 920470

Puzzle 34

```
2 9 7 9 9 6 1 7 0 4 7 9 3 9 8 4
4 1 9 7 0 3 1 0 7 0 9 3 4 9 6 6
3 8 3 1 6 9 9 5 4 7 4 1 2 7 0 7
0 0 7 1 6 1 0 6 8 0 4 6 6 0 5 2
7 8 1 1 9 9 9 3 3 8 2 5 9 2 7 9
5 9 4 9 1 8 2 3 1 3 7 5 2 7 8 2
0 8 2 6 8 7 7 1 7 5 8 4 4 7 7 3
1 3 3 0 4 8 5 0 8 5 6 6 4 5 8 7
6 6 2 7 7 1 9 7 7 0 3 1 4 5 2 2
3 1 0 3 1 5 6 7 2 7 1 0 5 6 7 3
9 2 8 6 6 9 9 1 7 9 2 1 5 9 3 9
4 7 3 0 9 1 3 4 4 7 1 3 5 6 7 8
7 6 7 1 0 1 3 5 3 8 0 2 5 6 9 6
6 2 0 8 5 0 5 3 3 9 9 7 7 2 1 4
3 6 1 7 1 1 8 9 8 9 8 0 4 6 0 8
5 1 2 2 1 5 1 0 4 1 5 3 0 4 5 4
8 1 5 1 1 8 7 7 9 0 9 2 9 8 5 0
9 8 9 9 2 4 0 8 2 5 7 7 4 0 7 6
```

- [ ] 161068
- [ ] 961383
- [ ] 093541
- [ ] 924445
- [ ] 419703
- [ ] 472787
- [ ] 655772
- [ ] 811518
- [ ] 960736
- [ ] 208370
- [ ] 723101
- [ ] 878750
- [ ] 613381
- [ ] 716997
- [ ] 471690
- [ ] 591101

Puzzle 35

```
0 1 6 2 5 9 3 2 7 9 8 8 2 4 4 8
2 0 7 1 2 2 9 8 7 1 4 8 6 0 4 5
0 1 3 1 6 8 1 1 9 6 2 7 3 6 4 7
6 1 0 5 2 6 7 5 8 3 5 4 2 4 3 9
0 8 3 5 4 8 3 0 2 4 4 2 8 6 7 6
1 8 7 5 7 8 0 2 7 5 7 6 6 6 5 5
4 5 6 6 7 7 8 2 2 7 2 7 9 7 7 5
7 8 5 2 8 5 3 8 8 2 4 9 4 4 2 7
4 5 6 7 6 4 9 8 4 0 5 1 7 4 2 1
2 4 0 5 3 5 9 0 8 5 4 0 6 1 0 4
4 6 9 8 0 5 7 1 0 2 0 6 7 8 0 0
9 5 5 4 5 3 1 1 8 1 7 7 4 2 3 5
7 8 3 7 8 5 0 5 2 5 3 9 8 4 6 7
9 1 3 6 7 1 6 6 9 3 0 5 3 4 2 1
8 4 7 2 4 8 4 9 6 2 0 8 4 2 6 7
0 8 6 1 6 1 3 7 4 5 0 9 9 0 4 5
4 7 5 8 7 7 0 2 2 7 3 8 8 5 7 3
4 1 6 4 3 7 6 6 8 0 6 8 4 6 6 4
```

- [ ] 001354
- [ ] 681196
- [ ] 275847
- [ ] 727228
- [ ] 785036
- [ ] 684918
- [ ] 854061
- [ ] 003548
- [ ] 029515
- [ ] 632869
- [ ] 564585
- [ ] 036264
- [ ] 354830
- [ ] 842547
- [ ] 359065
- [ ] 077857

Puzzle 36

```
3 3 8 0 7 8 1 9 6 7 7 3 4 7 9 4
0 9 3 4 5 0 4 9 8 6 0 6 0 2 2 1
6 7 2 6 0 2 2 3 9 1 6 7 4 8 0 2
4 9 3 6 9 6 8 0 9 2 3 2 6 5 7 6
3 8 5 6 7 8 2 7 0 1 5 6 9 5 2 1
5 1 5 8 9 8 4 5 7 8 1 2 6 2 6 8
9 5 7 0 8 6 9 2 0 9 7 8 0 0 0 6
2 7 9 4 6 7 1 5 5 0 7 1 3 6 4 3
1 5 6 5 8 3 4 7 4 8 5 0 0 0 9 9
4 0 3 4 5 0 0 0 1 0 3 2 3 8 8 7
1 6 4 3 7 5 0 1 4 8 2 8 0 6 1 7
7 8 2 6 5 7 0 7 4 0 6 6 0 4 7 3
5 8 0 4 5 7 0 9 2 5 2 3 9 2 7 5
2 7 6 8 8 5 5 9 8 2 1 0 2 8 1 5
6 8 5 9 4 9 3 1 5 3 3 8 8 1 5 6
6 4 4 9 3 7 8 0 3 6 1 5 6 0 0 5
7 7 8 2 5 8 9 7 2 2 5 2 3 0 3 3
3 2 1 0 4 1 4 5 8 8 4 1 6 9 7 5
```

- [ ] 012368
- [ ] 322700
- [ ] 377691
- [ ] 035564
- [ ] 608642
- [ ] 232908
- [ ] 920978
- [ ] 890555
- [ ] 385678
- [ ] 257141
- [ ] 255827
- [ ] 726022
- [ ] 517649
- [ ] 934504
- [ ] 685949
- [ ] 391674

Puzzle 37

```
6 9 3 2 3 9 3 1 4 0 6 1 8 6 7 4
6 8 3 2 8 9 0 1 9 4 0 0 9 1 8 7
3 0 9 8 9 9 5 6 0 3 4 4 7 4 2 9
9 4 9 7 4 7 4 4 7 4 0 1 4 8 2 2
8 2 0 3 7 1 6 6 2 7 1 9 7 3 7 5
0 2 0 7 7 4 3 0 4 5 4 4 9 0 0 6
7 1 6 1 1 5 9 9 0 9 5 0 0 8 1 8
4 0 6 4 0 3 9 2 9 0 9 8 9 9 5 8
4 1 0 4 3 0 1 5 7 1 2 3 3 1 8 1
5 0 3 2 2 8 5 8 5 4 5 3 9 2 8 5
8 8 1 0 4 1 5 3 4 4 6 7 8 5 9 9
0 0 7 6 2 0 7 9 3 4 4 9 2 1 3 5
4 7 0 5 5 4 4 3 0 1 8 6 3 6 3 1
3 8 0 7 8 4 1 9 1 3 4 8 9 1 3 7
6 2 1 1 1 0 7 2 6 3 9 5 2 1 5 4
8 3 5 2 4 6 3 6 2 3 9 3 8 5 8 5
3 0 0 2 8 9 9 8 3 7 4 7 5 9 3 4
7 3 7 2 7 0 2 9 2 7 5 8 3 4 7 3
```

- ☐ 859039 ☐ 599095 ☐ 899820 ☐ 980744
- ☐ 875803 ☐ 510722 ☐ 484652 ☐ 483089
- ☐ 160413 ☐ 289464 ☐ 067740 ☐ 544301
- ☐ 347702 ☐ 448131 ☐ 796853 ☐ 293289

Puzzle 38

```
5 5 2 9 1 5 1 7 9 7 4 1 7 2 1 0
7 4 6 3 5 1 9 2 1 0 2 0 1 4 2 5
6 8 3 4 2 8 4 8 3 6 2 1 8 0 5 0
5 3 8 3 0 7 0 3 6 0 2 0 4 7 6 6
5 2 5 0 8 4 7 6 2 2 2 0 4 7 3 5
9 9 7 5 5 4 2 4 5 7 7 8 7 7 1 9
9 8 2 8 7 8 3 9 1 3 6 0 1 1 9 8
9 0 5 7 9 1 4 6 5 5 5 5 1 8 4 3 1
3 1 4 8 0 9 3 2 3 7 0 2 0 8 7 0
4 4 9 1 2 6 1 2 3 3 9 9 9 0 8 7
0 0 5 5 2 1 4 9 5 8 7 9 0 4 3 3
5 5 6 8 4 7 9 4 3 8 9 2 5 7 5 4
9 3 1 9 2 1 7 0 4 4 5 6 1 1 8 5
1 8 1 2 7 7 8 3 8 6 4 2 5 4 5 6
8 2 7 4 5 7 0 8 7 6 1 1 0 5 3 2
9 4 0 4 1 1 6 4 6 5 5 5 5 2 9 1
0 4 6 8 4 2 7 6 5 3 3 3 6 6 0 0
4 3 7 1 3 4 5 4 3 5 4 9 1 5 8 4
```

- ☐ 836496 ☐ 097859 ☐ 858219 ☐ 327610
- ☐ 283504 ☐ 407129 ☐ 072343 ☐ 263857
- ☐ 099933 ☐ 817448 ☐ 573772 ☐ 333567
- ☐ 448196 ☐ 819504 ☐ 378358 ☐ 389257

Puzzle 39

```
4 7 9 4 7 6 5 5 2 6 1 3 6 9 6 7
3 4 5 4 7 3 7 5 3 1 8 3 9 1 2 3
4 2 4 7 5 5 9 3 8 1 6 8 8 1 2 2
5 4 2 5 2 1 9 3 5 9 8 0 0 5 1 8
6 5 1 1 7 6 4 1 5 9 2 2 5 9 6 7
0 6 8 7 1 2 3 0 4 8 7 5 1 1 0 8
3 6 8 3 6 2 8 3 1 6 6 2 9 9 7 8
7 7 0 6 7 1 5 9 2 0 0 9 1 0 1 0
7 4 7 3 9 3 1 8 1 1 4 2 5 0 8 6
0 4 5 4 4 0 2 9 2 1 5 6 3 4 7 2
2 2 0 7 8 0 8 8 7 6 2 4 8 7 2 2
7 6 7 2 5 0 5 8 1 8 7 8 0 4 9 6
7 5 8 1 2 4 7 9 0 8 9 6 6 1 8 5
5 6 3 2 9 6 7 6 4 1 9 5 5 7 0 1
6 8 3 0 3 5 4 0 8 7 5 9 6 8 0 5
8 1 5 7 3 9 6 1 5 6 9 2 7 6 4 2
0 4 1 3 7 7 7 1 2 0 5 0 8 0 6 8
7 8 1 5 0 8 5 5 7 5 7 2 7 7 6 2
```

- ☐ 052767 ☐ 808876 ☐ 806567 ☐ 479476
- ☐ 025292 ☐ 736347 ☐ 215634 ☐ 322112
- ☐ 451401 ☐ 706122 ☐ 566744 ☐ 123362
- ☐ 974218 ☐ 455832 ☐ 813573 ☐ 279959

Puzzle 40

```
6 8 8 3 8 0 9 6 8 1 4 5 0 4 5 1
0 5 0 2 5 2 4 2 8 6 8 1 8 7 5 8
6 4 4 8 4 1 8 2 1 7 0 7 0 6 8 1
9 7 6 5 9 2 4 2 1 4 3 5 2 2 0 0
9 4 8 5 1 3 6 9 5 3 9 3 4 0 1 3
5 1 0 3 8 8 3 1 1 7 7 8 4 1 8 7
6 6 3 4 8 0 0 6 4 2 4 1 2 7 8 2
9 4 0 6 2 5 6 7 7 9 5 4 6 4 2 8
5 5 9 2 1 0 8 4 3 9 4 3 9 6 4 1
4 2 3 8 2 1 1 4 9 9 7 9 9 2 7 2
9 1 5 9 5 1 3 3 8 5 1 6 1 9 3 8
3 2 2 9 5 2 3 0 8 5 6 7 4 0 2 9
4 7 8 2 5 5 5 4 2 6 8 6 6 1 3 8
2 8 0 3 8 7 7 3 4 5 9 8 4 8 3 6
8 7 4 1 6 9 2 9 3 4 2 5 9 0 0 4
3 5 6 2 3 6 6 2 6 2 6 5 8 7 2 6
2 7 6 2 0 4 6 3 4 8 0 8 2 0 7 0
3 6 7 5 5 4 6 5 4 1 1 9 9 6 0 5
```

- ☐ 361130 ☐ 602213 ☐ 843943 ☐ 583876
- ☐ 341242 ☐ 255542 ☐ 084364 ☐ 236626
- ☐ 952439 ☐ 442080 ☐ 513695 ☐ 418217
- ☐ 142460 ☐ 926471 ☐ 869083 ☐ 649852

Puzzle 41

```
9 8 1 7 6 0 6 5 7 4 2 0 2 5 0 9
0 8 1 3 1 5 6 1 9 3 4 6 3 6 2 4
4 0 3 8 0 7 5 8 4 7 4 6 1 5 9 8
1 4 4 2 0 9 4 6 0 9 6 9 1 5 8 8
3 9 1 6 0 8 9 9 2 0 0 8 6 9 9 9
9 6 7 4 9 4 4 2 4 2 3 3 6 0 2 7
3 0 2 9 0 4 6 6 5 8 6 1 9 7 3 0
8 7 6 1 6 7 6 1 4 1 1 3 7 4 7 1
3 0 4 5 5 7 1 6 8 5 8 6 7 9 9 8
6 2 4 0 0 4 2 0 7 4 6 7 4 4 1 1
3 0 6 4 5 7 9 9 7 8 4 6 0 4 8 2
8 8 2 9 6 2 8 3 4 3 8 4 3 0 7 2
2 7 1 1 6 9 1 5 2 2 7 6 1 3 2 6
9 1 8 7 4 9 9 0 2 2 2 6 7 4 7 4
6 2 5 1 7 9 3 4 0 5 1 2 6 7 8 3
6 2 0 5 7 9 9 7 3 8 6 9 9 3 1 5
0 2 4 2 1 4 3 2 0 9 5 8 9 3 2 9
0 4 0 2 6 6 2 0 1 5 0 6 1 1 7 0
```

- ☐ 420746
- ☐ 341242
- ☐ 491715
- ☐ 363839
- ☐ 200869
- ☐ 685867
- ☐ 167616
- ☐ 423360
- ☐ 152276
- ☐ 832046
- ☐ 784604
- ☐ 087122
- ☐ 474615
- ☐ 237918
- ☐ 916856
- ☐ 299031

Puzzle 42

```
1 5 9 6 1 1 3 5 2 2 4 1 7 1 4 7
8 8 0 0 1 9 9 8 7 1 4 1 8 1 5 7
4 8 8 5 5 8 5 2 6 2 5 8 0 4 7 6
9 7 6 5 9 4 3 7 3 0 7 9 0 3 0 2
4 3 0 9 8 4 1 2 1 9 4 2 8 9 9 7
9 7 9 1 3 9 2 4 5 9 2 8 1 7 7 1
8 2 2 0 7 5 1 5 2 8 5 3 8 4 3 6
7 4 1 7 6 2 0 0 9 3 5 1 2 2 0 3
3 3 8 3 7 0 8 9 1 7 9 0 5 4 3 1
4 5 2 1 4 3 5 9 2 8 8 9 1 2 3 8
8 0 1 1 3 3 5 6 2 8 2 6 9 2 1 3
1 2 0 5 6 2 2 7 5 8 1 3 2 5 2 4
5 9 3 3 0 2 5 6 7 7 9 1 5 2 2 3
4 6 9 5 5 4 5 6 5 9 3 9 0 4 6 9
7 4 4 2 2 9 0 0 2 6 8 4 3 9 9 8
3 2 3 1 6 1 8 0 6 5 5 0 6 2 5 9
4 3 8 5 2 2 5 1 7 3 1 4 1 7 0 3
2 2 5 7 5 0 9 5 7 9 6 1 6 2 3 6
```

- ☐ 127275
- ☐ 634767
- ☐ 495679
- ☐ 978622
- ☐ 326565
- ☐ 349301
- ☐ 764375
- ☐ 625855
- ☐ 033122
- ☐ 378951
- ☐ 931979
- ☐ 214890
- ☐ 328101
- ☐ 481547
- ☐ 515917
- ☐ 522517

Puzzle 43

```
1 3 4 6 1 2 8 0 2 7 4 3 0 0 9 4
8 2 0 5 5 8 7 4 1 0 3 2 7 6 7 9
7 3 9 7 7 4 9 7 3 0 8 1 9 7 0 2
1 5 4 8 7 8 5 3 7 2 7 9 2 7 9 6
0 5 8 5 5 0 2 4 4 4 1 8 0 6 5 3
6 1 8 0 7 9 9 8 2 7 3 4 3 3 8 6
1 2 8 8 9 5 1 4 3 0 2 8 2 2 4 6
3 9 6 1 6 2 0 7 8 3 4 4 8 9 0 4
3 8 4 2 1 1 9 4 0 3 4 4 5 7 1 2
5 2 1 3 6 7 0 9 0 0 4 9 7 7 9 8
0 4 1 8 2 9 9 1 4 0 2 6 9 9 1 9
8 2 4 1 7 5 4 3 8 1 2 1 8 3 4 0
5 1 4 4 1 2 3 5 3 8 0 0 6 4 6 6
9 5 5 3 8 6 5 6 5 4 8 3 5 4 7 7
1 6 9 4 7 9 5 9 7 5 7 0 7 8 6 8
7 9 0 9 3 7 6 6 0 7 0 7 6 1 3 4
1 5 5 9 3 2 6 6 0 2 8 6 3 2 4 2
2 3 7 4 4 2 1 8 2 7 5 1 4 4 5 5
```

- ☐ 081444
- ☐ 016885
- ☐ 908482
- ☐ 703300
- ☐ 834571
- ☐ 698444
- ☐ 783448
- ☐ 719580
- ☐ 301478
- ☐ 410371
- ☐ 677342
- ☐ 538006
- ☐ 992908
- ☐ 660707
- ☐ 711810
- ☐ 397792

Puzzle 44

```
4 4 1 1 6 8 5 6 2 3 0 4 6 8 3 6
1 1 9 6 9 6 2 2 6 6 3 3 4 7 3 5
6 8 9 8 3 4 7 8 9 6 3 4 5 7 0 8
8 3 6 9 1 5 2 0 5 2 1 6 4 8 4 7
0 4 7 1 3 3 4 7 2 5 1 6 4 3 3 4
3 4 4 7 6 3 0 5 4 0 1 5 7 6 9 7
6 1 8 4 1 8 5 0 7 7 9 7 5 3 5 2
5 4 3 4 6 0 7 6 0 4 5 9 4 5 0 7
7 9 9 5 3 2 0 7 6 0 8 6 4 6 9 5
8 2 1 3 9 5 3 4 9 9 6 4 3 5 9 1
3 1 7 6 5 5 9 6 6 7 6 1 0 4 4 4
7 1 0 7 8 1 3 6 5 0 4 5 0 6 3 4 9
3 4 3 4 9 6 0 6 9 7 6 4 9 0 1 8
0 5 9 1 8 6 3 1 9 7 7 3 4 1 9 5
4 3 4 3 5 3 5 8 7 2 5 2 3 8 8 9
7 0 9 6 8 8 1 3 5 4 5 5 3 9 7 3
1 0 6 8 5 5 3 3 8 6 6 0 9 8 0 4
8 3 0 8 0 2 3 0 0 6 2 0 9 9 5 1
```

- ☐ 985936
- ☐ 003189
- ☐ 579906
- ☐ 386403
- ☐ 315303
- ☐ 653399
- ☐ 779754
- ☐ 336622
- ☐ 354574
- ☐ 583661
- ☐ 574724
- ☐ 093364
- ☐ 454475
- ☐ 944198
- ☐ 418344
- ☐ 043950

Puzzle 45

```
8 1 1 9 3 7 6 4 1 8 4 6 3 3 8 2
1 1 2 1 1 4 2 0 1 8 4 5 1 9 9 1
6 6 1 8 0 4 9 3 5 1 1 7 6 6 2 1
7 0 8 2 5 7 4 0 3 9 2 8 7 0 8 6
7 5 9 1 1 7 6 6 8 3 0 2 3 3 2 5
7 2 9 5 5 1 6 0 8 4 7 8 7 3 2 0
3 5 4 6 2 3 6 9 6 4 4 8 1 3 6 8
5 2 2 0 9 2 7 2 7 8 0 4 1 0 7 5
7 9 1 1 1 9 6 2 9 0 0 0 6 1 2 5
0 7 6 5 2 4 2 4 2 5 5 8 6 0 3 7
8 1 4 7 3 5 6 9 3 4 0 2 6 8 5 1
2 7 9 8 4 6 7 7 3 7 8 5 0 3 0 6
1 1 6 7 6 3 7 3 4 5 5 5 4 0 8 8
2 6 8 8 2 8 9 6 9 4 8 4 4 3 4 0
2 0 6 5 6 8 2 6 7 3 0 7 4 0 6 9
7 5 3 4 6 3 3 4 6 8 9 4 9 0 8 7
5 7 0 9 7 2 9 5 1 2 0 8 3 4 2 6
1 6 0 8 9 3 7 7 8 4 7 2 1 4 6 2
```

- [ ] 984677
- [ ] 759117
- [ ] 673711
- [ ] 908444
- [ ] 097295
- [ ] 144684
- [ ] 830300
- [ ] 629000
- [ ] 085805
- [ ] 242425
- [ ] 065128
- [ ] 935117
- [ ] 243802
- [ ] 653741
- [ ] 206568
- [ ] 672350

Puzzle 46

```
5 4 3 9 0 0 1 3 0 7 7 2 2 6 2 7
1 4 0 3 6 6 6 6 3 4 5 3 0 1 1 8
6 8 1 6 7 7 5 6 3 4 8 8 1 7 8 3
6 7 5 1 8 1 6 8 9 5 7 1 8 9 9 0
3 4 5 7 5 5 6 9 1 1 8 6 3 5 5 7
8 8 0 8 8 2 2 4 4 9 4 8 0 7 1 5
2 2 1 4 3 3 8 4 7 4 2 4 9 8 8 8
0 5 0 4 0 9 1 2 8 1 5 9 8 4 4 6
0 2 6 5 6 9 9 9 6 3 9 1 0 1 0 3 8
9 7 7 3 5 9 2 4 9 4 9 3 2 6 4 1
1 5 2 8 0 6 3 0 2 4 5 6 3 6 5 1
0 9 7 4 3 8 0 7 8 5 0 2 3 7 2 2
8 7 7 0 5 2 6 5 1 7 9 8 2 5 1 7
8 2 9 4 1 5 6 2 5 6 2 5 8 1 5 8
7 4 8 4 5 5 7 0 6 1 8 1 7 8 1 1 1
5 8 4 1 8 9 8 6 2 0 2 2 9 8 3 4
0 6 5 2 1 9 6 1 8 2 7 4 0 8 1 6
6 5 6 9 1 2 8 3 8 7 3 8 9 7 7 6
```

- [ ] 965575
- [ ] 381469
- [ ] 675118
- [ ] 710759
- [ ] 543282
- [ ] 291826
- [ ] 889424
- [ ] 807850
- [ ] 100934
- [ ] 256258
- [ ] 226623
- [ ] 724550
- [ ] 675181
- [ ] 912560
- [ ] 819291
- [ ] 577618

Puzzle 47

```
4 5 9 3 2 1 2 1 6 0 4 0 8 2 0 4
5 3 0 3 7 2 6 6 4 4 3 8 9 6 4 7
1 0 6 1 8 7 9 6 7 6 2 7 4 3 0 6
9 4 3 6 3 8 6 2 2 3 3 1 0 8 6 2
9 7 1 1 4 9 4 0 0 6 4 1 8 9 9 9
0 2 7 8 4 9 2 6 6 7 5 8 0 1 2 5
2 7 6 8 7 2 3 2 0 6 4 0 8 8 3 0
6 8 1 0 5 7 2 1 0 9 0 0 4 9 5 7
0 1 0 7 0 8 3 9 9 2 4 9 4 9 4 2
6 7 6 9 9 2 6 4 3 4 8 3 8 5 9 0
2 0 6 9 2 3 9 6 4 3 0 9 2 0 7 2
4 5 6 3 9 0 4 3 1 9 9 5 4 1 9 3
9 4 4 6 8 5 9 6 4 9 1 7 3 7 1 9
9 2 5 7 0 8 1 0 5 8 1 8 8 7 4 8
1 8 9 2 7 7 3 2 3 4 0 5 3 7 1
3 6 7 3 8 8 0 0 5 2 0 5 9 3 2 7
6 6 0 6 3 7 1 6 6 3 7 9 9 5 1 2
0 8 1 6 3 1 5 3 7 0 3 6 6 7 7 9
```

- [ ] 982022
- [ ] 081058
- [ ] 194636
- [ ] 844824
- [ ] 389189
- [ ] 997088
- [ ] 061879
- [ ] 491853
- [ ] 409365
- [ ] 711494
- [ ] 454323
- [ ] 238793
- [ ] 638622
- [ ] 377606
- [ ] 994260
- [ ] 940537

Puzzle 48

```
7 2 9 5 2 4 7 9 1 4 6 7 3 2 6 7
4 2 5 7 2 6 1 2 5 8 0 8 5 0 7 6
7 0 1 3 5 6 4 7 9 4 1 9 0 4 9 1
5 0 3 0 3 3 1 4 6 8 9 9 1 3 8 9
2 3 6 4 5 3 4 8 4 9 2 7 4 1 0 3
7 0 6 7 0 4 7 2 0 2 6 3 2 9 5 2
8 0 3 5 1 5 9 2 7 1 8 5 3 3 6 9
0 5 9 6 3 7 8 4 1 6 4 2 4 9 0 2
8 6 4 0 1 9 1 9 4 8 3 8 0 3 2 0
8 4 7 2 2 7 8 7 3 6 0 2 2 0 2 5
6 0 8 8 1 7 4 6 0 5 2 1 0 3 5 5
7 5 2 8 7 9 1 5 9 4 4 3 7 3 6 7
8 8 4 2 7 0 0 9 9 8 5 8 0 7 5 3
4 7 8 6 3 8 7 4 4 7 3 2 7 1 8 0
1 3 2 0 0 7 7 5 4 1 1 1 8 4 4 3
6 0 3 8 4 3 2 8 8 0 2 7 0 8 4 1
3 6 3 0 1 0 2 6 1 2 7 4 9 5 1 4
4 0 1 6 9 5 5 8 1 5 9 7 1 3 4 2
```

- [ ] 088174
- [ ] 391340
- [ ] 784539
- [ ] 358172
- [ ] 565220
- [ ] 372180
- [ ] 786387
- [ ] 721620
- [ ] 350142
- [ ] 114577
- [ ] 484927
- [ ] 125808
- [ ] 067171
- [ ] 867841
- [ ] 135647
- [ ] 177304

Puzzle 49

```
4 2 5 9 8 5 7 7 4 7 0 3 0 9 1 9
7 0 5 7 4 3 4 2 0 3 4 1 4 0 3 1
4 4 5 9 9 9 3 9 9 9 3 7 8 9 1 5
9 5 9 6 3 2 3 6 7 1 4 5 4 8 3 5
9 9 0 5 7 1 2 9 6 4 0 5 3 5 8 5
8 8 1 8 9 3 9 6 3 1 5 4 5 8 1 2
0 0 9 6 9 2 6 0 4 6 2 5 2 4 6 7
0 4 1 9 1 7 1 4 7 0 4 7 2 9 4 5
2 8 9 6 3 1 0 8 1 1 5 3 0 1 2 5
8 8 3 4 2 2 8 2 9 3 8 8 7 6 2 0
1 2 3 8 2 3 2 4 9 2 0 1 8 7 3 3
6 3 3 1 7 3 0 1 6 9 6 2 6 9 1 5
0 5 0 9 3 7 5 1 5 1 0 0 7 4 2 5
2 0 5 6 0 3 7 5 2 6 3 2 2 4 7 2
0 7 5 3 6 0 6 9 4 0 8 1 5 0 4 0
9 7 0 8 6 5 1 2 6 9 1 2 3 9 4 8
0 7 3 6 6 5 4 4 5 0 9 4 8 6 0 6
2 0 3 1 4 6 1 4 1 8 2 7 8 7 0 4
```

- ☐ 369398
- ☐ 640629
- ☐ 399954
- ☐ 069408
- ☐ 823507
- ☐ 060854
- ☐ 037526
- ☐ 381202
- ☐ 631316
- ☐ 420341
- ☐ 207867
- ☐ 532494
- ☐ 044976
- ☐ 989932
- ☐ 296405
- ☐ 714548

Puzzle 50

```
4 5 7 7 9 6 9 2 3 3 5 9 7 0 7 8
7 6 5 6 0 1 7 8 6 4 0 0 7 3 7 6
1 3 4 7 6 0 0 7 1 1 1 7 7 7 5 0
8 1 4 5 5 1 9 5 3 1 4 8 6 8 0 7
3 8 2 5 3 7 9 5 8 6 6 1 7 9 9 6
1 8 2 9 9 7 3 3 3 8 1 3 6 7 1 5
0 8 2 2 6 9 5 3 1 2 4 1 6 1 6 7
8 5 0 3 3 4 0 7 9 1 1 1 0 9 1 8
3 5 9 7 4 8 7 7 9 0 6 5 0 2 5 7
9 9 7 3 9 5 0 6 7 8 2 0 2 1 4 2
5 0 5 9 1 3 1 4 8 8 6 3 3 6 8 3
3 8 5 8 6 8 5 3 7 1 5 9 8 0 4 7
8 9 9 3 2 9 4 4 3 5 9 0 5 6 1 2
3 4 6 8 7 7 7 8 6 8 4 5 9 7 5 2
8 8 3 1 4 2 4 5 9 2 2 5 7 7 4 0
0 6 6 7 2 5 4 3 9 4 2 7 8 4 3 8
1 6 0 7 6 7 0 5 0 5 0 3 6 8 1 9
5 6 3 5 4 3 3 9 8 6 0 2 4 1 5 3
```

- ☐ 757339
- ☐ 332969
- ☐ 235900
- ☐ 905877
- ☐ 361102
- ☐ 319505
- ☐ 574083
- ☐ 589845
- ☐ 636118
- ☐ 631888
- ☐ 879583
- ☐ 584346
- ☐ 577711
- ☐ 391667
- ☐ 516190
- ☐ 606774

Puzzle 51

```
6 4 3 7 3 6 4 5 9 3 6 0 4 5 6 0
0 8 5 2 5 6 3 0 9 8 1 5 7 8 1 0
2 4 5 4 6 4 5 5 7 1 0 3 8 0 1 5
2 2 4 6 7 2 2 7 2 5 0 8 6 0 0 8
0 2 8 6 1 8 0 9 2 7 0 9 5 1 9 4
8 6 6 9 3 1 4 6 3 8 1 0 6 8 4 5
5 7 2 4 7 8 3 3 5 0 0 6 7 3 5 8
2 1 5 7 0 5 4 2 8 4 3 9 9 0 6 3
7 5 9 8 6 5 9 5 1 7 4 5 3 5 9 8
5 0 0 5 3 3 3 5 6 0 2 0 0 2 7 3
7 3 0 1 3 7 9 5 6 4 8 5 0 8 8 8
6 3 7 9 5 8 2 7 3 2 2 0 8 4 7 1
9 5 1 3 3 5 0 6 6 7 4 6 8 2 0 2
6 9 4 0 2 9 2 9 3 7 7 1 9 1 9 8
6 8 7 6 5 3 9 9 7 8 4 7 6 6 8 7
9 6 4 5 1 8 3 4 2 1 4 8 7 3 7 0
6 7 6 5 4 2 0 9 4 9 5 0 5 4 0 6
1 0 8 3 8 0 7 3 0 4 2 0 7 1 6 1
```

- ☐ 624654
- ☐ 179083
- ☐ 818246
- ☐ 265957
- ☐ 823908
- ☐ 535377
- ☐ 737841
- ☐ 098157
- ☐ 303508
- ☐ 700939
- ☐ 527169
- ☐ 255105
- ☐ 762248
- ☐ 537323
- ☐ 800183
- ☐ 736459

Puzzle 52

```
9 1 0 4 2 3 6 4 9 7 9 4 7 1 7 9
6 0 5 7 2 5 7 9 7 6 8 1 9 2 0 8
6 3 1 2 9 8 7 0 0 8 0 6 6 7 0 7
2 1 5 0 5 6 8 9 0 1 5 5 8 2 3 7
0 6 2 9 2 7 8 4 7 1 4 0 7 3 6 4
1 2 1 5 7 2 0 9 7 5 6 0 7 5 8 5
8 4 4 3 5 7 5 1 1 0 7 8 3 3 4 2
3 2 2 0 7 7 6 1 9 1 6 8 5 8 6 7
7 0 6 0 4 3 7 1 0 9 6 9 6 9 6 1
9 1 0 9 2 7 2 2 9 3 6 5 3 9 6 4
3 6 8 4 4 6 5 2 6 4 0 9 0 0 8 0
1 3 5 0 7 9 5 2 3 9 4 7 0 9 9 9
7 7 6 6 8 1 3 9 6 6 6 3 1 0 2 3
1 9 6 2 0 9 0 6 8 7 6 2 5 2 6 3
7 3 9 7 4 6 0 8 9 4 2 5 2 4 2 1
5 0 0 7 0 6 7 3 5 2 8 2 8 9 2 4
9 0 1 7 7 4 2 2 3 4 9 9 8 9 9 6
6 6 0 2 1 1 1 6 9 8 0 9 5 6 7 0
```

- ☐ 252678
- ☐ 903960
- ☐ 672773
- ☐ 631298
- ☐ 900359
- ☐ 656067
- ☐ 749794
- ☐ 008066
- ☐ 116509
- ☐ 901774
- ☐ 714073
- ☐ 317175
- ☐ 018379
- ☐ 636909
- ☐ 966631
- ☐ 725922

Puzzle 53

```
0 3 4 4 2 4 8 0 2 8 9 0 3 1 9 8
4 7 6 6 3 6 5 2 9 4 5 5 5 0 2 4
7 1 3 9 8 5 2 8 9 5 5 8 3 3 2 1
2 1 2 0 6 1 1 6 1 4 8 6 7 2 6 4
1 1 1 1 0 6 1 0 8 8 2 8 1 5 7 4
8 7 6 6 5 2 3 0 4 5 5 0 8 8 4 9
6 9 7 3 3 9 1 0 3 4 8 0 6 4 1 5
9 2 8 6 9 9 6 1 5 5 2 4 9 5 1 2
7 6 8 3 0 4 8 8 3 6 5 5 5 5 0 3
9 6 8 6 6 9 8 2 8 0 6 2 7 3 9 6
8 6 1 6 0 4 7 1 1 6 8 8 9 9 7 5
4 3 3 7 7 4 9 9 6 3 3 6 1 7 3 1
1 1 9 8 6 0 3 6 8 0 3 2 0 3 4 4
3 7 8 2 3 4 9 1 7 2 6 5 9 5 7 4
9 0 8 4 8 3 0 8 4 8 6 5 3 4 4 6
4 8 1 6 2 6 6 0 7 5 9 0 2 1 6 9
1 5 6 3 8 2 7 7 4 8 2 3 5 9 4 9
7 6 5 2 0 8 9 9 1 9 4 1 6 7 1 3
```

☐ 845485	☐ 726595	☐ 476636	☐ 219618
☐ 106108	☐ 011476	☐ 998861	☐ 356848
☐ 032584	☐ 665230	☐ 866982	☐ 193379
☐ 337749	☐ 748235	☐ 048836	☐ 558258

Puzzle 54

```
6 9 2 9 9 1 7 9 8 3 0 5 0 3 1 7
3 9 2 8 2 3 3 7 6 3 1 0 2 5 3 7
5 7 5 0 7 2 6 0 5 1 4 9 4 8 6 7
7 9 0 6 9 4 0 2 4 0 6 5 2 4 9 1
2 3 9 7 5 9 4 4 7 1 1 8 2 0 5 4
7 4 4 3 2 2 3 4 9 3 4 8 5 2 1 0
4 2 4 6 5 6 4 0 5 3 4 0 6 6 9 9
8 0 3 4 7 4 0 9 4 9 7 5 9 2 4 5
2 1 2 9 6 2 4 5 7 7 9 0 8 6 1 9
9 8 1 2 5 0 8 5 6 1 6 9 2 1 7 5
4 1 2 1 8 2 1 2 4 7 0 9 2 6 3 5
6 8 1 5 8 4 8 0 6 6 9 8 4 3 3 9
5 6 8 6 7 1 1 0 3 6 3 0 5 8 3 9
0 9 3 1 0 5 7 5 1 3 7 4 6 6 0 9
2 6 3 5 0 4 4 0 1 8 6 6 8 7 9 2
3 5 0 4 5 6 3 1 3 0 8 6 1 6 1 4
8 6 1 4 3 4 9 1 7 6 1 3 5 5 5 5
8 7 5 9 3 9 4 2 8 7 0 7 2 5 6 9
```

☐ 575072	☐ 798305	☐ 865422	☐ 631025
☐ 904743	☐ 225094	☐ 658052	☐ 624577
☐ 616262	☐ 928472	☐ 849415	☐ 960188
☐ 273458	☐ 616803	☐ 322349	☐ 951941

Puzzle 55

```
6 7 2 9 5 4 2 3 2 5 5 3 6 5 8 7
7 4 6 5 9 8 8 4 2 4 0 8 0 1 9 8
4 5 3 7 0 5 4 4 0 2 0 7 4 9 0 9
3 8 1 8 7 5 2 5 5 6 5 0 6 6 1 6
8 7 2 2 0 2 9 2 9 7 6 4 5 7 9 6
4 4 2 0 3 1 7 2 3 0 0 7 0 8 8 9
9 9 5 5 5 4 1 7 8 2 6 8 3 5 2 5
8 1 7 8 9 4 8 5 3 0 6 3 6 1 4 5
5 0 7 3 1 3 3 7 3 5 8 3 9 0 5 1
2 6 8 6 1 3 5 5 3 0 2 7 6 2 6 6
7 4 2 1 3 8 4 4 6 3 4 9 7 0 6 4
5 3 4 9 4 1 6 5 7 1 7 4 0 9 0 2
0 6 1 3 6 3 6 4 4 3 9 0 3 9 8 9
1 4 2 4 7 1 9 2 5 7 6 7 0 0 7 6
0 1 8 6 1 5 8 8 3 8 9 0 3 3 4 3
7 5 6 8 9 4 8 7 2 6 9 6 9 8 0 2
9 0 6 9 5 2 0 0 5 1 1 7 5 2 0 7
8 9 5 3 1 2 6 5 4 3 2 4 5 9 3 2
```

☐ 636035	☐ 529174	☐ 507313	☐ 348733
☐ 634970	☐ 095485	☐ 296265	☐ 952232
☐ 428910	☐ 726969	☐ 005117	☐ 631225
☐ 967851	☐ 804248	☐ 439039	☐ 571740

Puzzle 56

```
9 9 0 4 2 9 9 6 1 5 0 0 3 2 3 1
9 5 5 7 1 8 3 3 1 6 7 1 7 7 9 2
6 9 4 0 0 3 6 1 0 5 9 5 8 1 7 0
8 8 5 4 0 0 6 3 4 5 9 3 4 7 2 0
8 0 4 3 0 9 6 0 1 6 3 4 8 2 0 5
1 3 1 0 3 7 9 1 9 5 8 8 2 2 4 9
1 2 3 5 4 2 3 1 5 9 7 3 7 6 7 5
1 6 4 0 5 3 0 2 5 8 9 6 7 3 8 7
3 5 6 2 4 4 4 5 6 0 9 7 5 7 5 8
8 7 3 1 3 9 6 1 6 1 7 2 0 7 8 3
1 0 7 5 5 3 7 0 4 4 0 8 3 0 5 7
8 8 7 8 1 2 7 2 2 1 4 3 8 8 8 3
0 0 1 9 3 5 8 9 5 3 5 6 0 9 0 1
3 2 6 2 7 3 7 3 0 6 9 2 7 7 7 9
3 6 7 9 4 6 8 8 2 5 3 1 0 6 8 8
4 9 4 9 6 4 3 2 5 3 4 6 5 7 1 2
8 9 8 9 8 1 9 6 0 2 5 7 0 3 9 5
3 2 5 3 4 4 7 8 4 0 4 4 1 4 8 4
```

☐ 276384	☐ 544073	☐ 852035	☐ 688111
☐ 562444	☐ 044083	☐ 469494	☐ 946882
☐ 272187	☐ 141365	☐ 784827	☐ 966639
☐ 798077	☐ 602570	☐ 432534	☐ 693137

Puzzle 57

```
8 3 4 3 8 9 1 4 6 7 3 2 0 8 2 1
2 6 6 5 3 0 0 2 2 0 3 3 9 1 4 7
1 8 8 6 0 6 1 5 9 5 8 6 0 9 2 7
1 0 7 9 6 3 6 2 9 9 0 5 5 4 5 9
2 0 0 4 1 6 6 7 2 9 5 8 6 3 6 7
0 2 1 8 0 7 7 9 6 6 7 4 7 5 1 1
6 5 2 3 7 0 4 4 5 3 4 7 6 5 5 2
6 9 4 9 4 6 4 8 9 7 1 3 1 4 3 4
9 1 7 3 1 6 5 0 2 8 4 9 9 5 5 8
3 8 9 7 7 9 4 1 4 8 8 9 2 3 6 5
7 7 7 2 5 2 7 5 0 8 2 2 1 5 0 3
6 3 8 4 6 0 3 4 0 3 5 4 1 4 4 0
7 3 5 1 7 7 8 7 9 3 4 6 9 1 2 4
4 4 8 4 6 5 1 4 2 0 1 4 7 8 6 2
1 6 6 7 7 8 8 7 2 1 2 4 4 1 2 2
4 2 6 7 9 4 6 3 0 9 7 0 8 3 5 6
9 4 0 8 8 0 8 5 5 8 7 3 5 4 8 2
5 3 4 6 0 1 5 1 9 6 7 8 8 6 4 9
```

☐ 100876	☐ 636299	☐ 433781	☐ 527508
☐ 318145	☐ 250875	☐ 004782	☐ 427393
☐ 763192	☐ 071762	☐ 464599	☐ 855873
☐ 749882	☐ 604262	☐ 146732	☐ 369574

Puzzle 58

```
3 8 2 5 8 4 6 8 5 9 5 8 0 4 7 4
3 1 3 4 4 2 0 7 6 2 7 1 3 8 4 8
8 3 3 8 9 5 4 9 6 1 1 4 2 7 2 3
3 1 4 3 0 7 0 5 3 3 9 7 2 0 9 6
5 9 0 7 6 7 3 6 1 6 3 0 9 1 5 2
6 4 7 1 7 1 6 6 0 1 1 1 4 6 8 8
0 9 3 4 4 5 2 7 6 9 0 7 3 9 4 3
5 3 0 0 9 2 1 6 6 4 1 6 0 3 2 2
9 4 7 0 1 8 8 6 1 8 3 3 5 5 2 8
9 2 7 6 0 8 0 4 4 1 7 2 9 4 7 1
5 4 6 5 2 0 8 9 9 3 0 8 8 0 0 9
0 5 7 6 9 3 2 7 9 6 9 1 1 5 1 0
6 5 0 7 6 5 1 0 9 7 2 1 4 2 8 7
8 9 5 9 8 8 4 0 3 0 5 3 6 9 7 4
3 4 6 1 5 3 1 4 7 2 0 3 8 1 6 1
5 2 7 1 0 7 8 3 3 8 4 6 2 4 1 0
0 6 7 9 1 7 1 6 4 6 2 3 1 0 0 8
7 1 1 6 0 3 2 0 4 9 9 6 5 4 3 9
```

☐ 118236	☐ 949131	☐ 416601	☐ 472038
☐ 080672	☐ 079466	☐ 694598	☐ 995065
☐ 507651	☐ 076318	☐ 493709	☐ 925045
☐ 326461	☐ 793350	☐ 701725	☐ 771528

Puzzle 59

```
5 6 2 4 4 1 8 0 8 2 1 9 5 0 3 5
9 1 5 7 8 0 5 1 8 4 0 4 2 9 7 0
1 4 6 9 1 7 0 3 7 7 1 3 4 1 3 6
2 3 1 9 9 6 0 3 2 1 7 2 3 5 9 9
9 8 6 6 1 9 3 0 0 8 8 8 8 9 5 4
6 8 8 6 6 8 5 5 1 7 6 6 5 7 2 6
0 0 7 7 8 8 7 7 9 1 4 5 6 0 3 5
9 0 0 2 5 5 5 5 8 1 6 5 5 2 9 8
1 5 7 6 7 3 3 9 6 0 7 7 3 0 9 5
7 1 1 1 8 7 7 4 4 2 4 0 1 6 0 7
0 9 4 7 8 2 9 0 1 8 3 8 0 3 9 5
4 4 1 7 3 2 3 0 1 8 8 1 1 8 6 2
5 8 4 6 5 5 1 6 3 5 4 3 5 6 7 0
0 8 5 7 8 9 9 2 5 4 4 1 5 1 4 5
9 8 1 1 1 5 3 5 9 9 0 0 8 0 3 0
6 4 0 7 0 2 0 1 8 3 6 1 4 0 1 1
1 3 0 6 4 6 6 4 5 1 3 0 7 7 5 6
2 2 8 8 0 7 4 0 0 7 3 9 6 1 0 2
```

☐ 787700	☐ 410288	☐ 619184	☐ 756682
☐ 621744	☐ 518404	☐ 043097	☐ 379519
☐ 481435	☐ 948835	☐ 185953	☐ 351118
☐ 058342	☐ 037713	☐ 134769	☐ 094782

Puzzle 60

```
7 5 5 4 9 8 9 6 7 2 4 4 0 9 4 5
2 0 6 2 4 8 4 4 2 4 7 5 0 5 3 4
1 8 7 0 9 8 3 4 4 3 0 9 7 5 1 6
7 0 7 7 7 4 6 9 3 6 4 8 7 7 0 7
4 8 8 6 0 5 1 3 0 8 4 5 3 9 8 7
2 0 8 6 1 7 6 8 9 4 7 4 6 8 4 6
8 9 2 2 5 5 4 2 3 5 3 8 4 0 3 4
5 6 6 2 6 6 5 8 2 0 1 1 9 6 4 5
0 6 1 8 2 1 7 4 0 4 0 6 7 8 1 9
7 4 9 9 5 0 5 6 0 6 3 4 1 4 0 9
0 4 9 9 4 4 4 5 1 9 2 4 5 0 8 1
8 3 3 7 7 0 2 2 0 2 1 0 6 6 5 0
2 5 1 7 9 4 2 5 2 3 9 0 2 3 7 7
5 4 9 7 9 3 5 9 2 1 7 1 0 8 6 4
0 0 7 7 2 3 6 5 3 6 6 4 9 0 1 7
4 9 8 7 6 4 0 8 3 4 1 5 7 7 6 1
1 5 2 3 9 6 0 8 6 2 4 5 3 9 2 2
9 5 5 5 3 6 2 2 1 1 6 7 1 7 1 4
```

☐ 466908	☐ 220773	☐ 814925	☐ 427130
☐ 014348	☐ 002390	☐ 612919	☐ 501615
☐ 470447	☐ 156254	☐ 217428	☐ 985481
☐ 077074	☐ 341577	☐ 806932	☐ 794637

Puzzle 61

```
9 0 9 6 9 1 9 6 6 1 2 8 2 2 3 1
8 5 2 6 5 7 1 5 7 9 3 3 2 7 8 5
6 6 8 7 7 4 9 2 7 0 1 5 7 5 1 3
7 6 5 0 8 9 1 5 4 7 3 3 5 0 7 3
0 5 1 1 2 5 2 0 0 7 8 5 9 7 4 3
9 7 2 2 7 8 1 6 1 1 1 4 3 6 5 5
3 5 2 9 2 3 6 1 9 9 3 0 1 4 0 2
2 7 3 7 1 6 7 2 7 2 7 3 8 7 1 1
0 4 5 9 4 8 1 2 3 0 2 7 5 3 8 3
7 7 3 0 7 6 0 3 8 6 1 7 9 1 8 9
3 5 5 8 9 0 8 5 9 9 1 7 3 1 3 9
3 8 3 1 2 5 6 5 2 5 0 1 0 5 9 5
1 8 0 6 6 7 2 9 2 7 0 1 4 6 5 6
5 2 6 6 5 7 1 7 7 9 9 9 7 6 2 9
0 5 7 3 0 7 4 2 6 2 5 8 6 6 2 2
1 0 4 0 8 7 2 3 9 1 5 8 9 5 9 2
0 4 2 5 7 1 5 7 8 1 1 0 5 1 6 1
1 6 6 9 6 5 6 2 5 2 8 1 2 2 4 0
```

- [ ] 989725
- [ ] 297908
- [ ] 353540
- [ ] 926508
- [ ] 071158
- [ ] 501187
- [ ] 656905
- [ ] 368605
- [ ] 023907
- [ ] 225938
- [ ] 779101
- [ ] 574758
- [ ] 216691
- [ ] 593047
- [ ] 792692
- [ ] 175240

Puzzle 62

```
4 2 5 9 8 8 5 8 0 3 4 7 8 1 5 8
3 5 6 6 4 1 4 3 8 1 6 7 9 0 8 2
7 3 2 8 2 8 1 7 8 1 1 8 2 5 6 6
0 4 0 5 6 0 9 7 1 5 3 0 6 9 2 3
7 1 4 4 1 6 4 2 4 1 6 9 1 4 2 1
7 8 9 4 2 8 2 2 6 9 0 3 6 3 9 7
0 7 3 6 0 1 5 5 5 0 9 5 2 2 6 2
8 4 2 4 3 5 8 2 3 8 8 8 7 8 2 8
5 3 6 7 1 7 3 9 0 6 6 4 4 1 8 7
5 5 7 3 4 1 1 3 3 4 1 7 8 1 0 5
5 2 1 8 6 1 7 8 9 1 6 1 3 3 4 3
2 7 2 4 4 8 7 2 1 9 0 7 9 5 2 1
2 0 1 1 1 5 3 3 4 4 6 6 1 8 2 8
7 0 3 8 4 3 6 8 2 2 5 2 1 7 5 4
9 6 3 1 1 8 1 0 3 9 9 3 9 2 8 6
2 3 4 2 2 9 9 1 9 7 0 2 2 8 6 2
4 5 0 5 4 1 0 0 1 1 9 7 3 6 8 7
4 2 3 2 5 3 1 1 1 5 0 5 0 8 3 5
```

- [ ] 750117
- [ ] 489947
- [ ] 328534
- [ ] 425867
- [ ] 131415
- [ ] 095606
- [ ] 311823
- [ ] 261627
- [ ] 360155
- [ ] 430858
- [ ] 912784
- [ ] 221489
- [ ] 555227
- [ ] 807707
- [ ] 848062
- [ ] 183414

Puzzle 63

```
5 5 7 7 9 8 9 9 4 1 8 0 9 3 4 3
0 3 9 0 3 1 3 3 6 3 3 0 8 1 6 1
9 9 1 0 8 2 7 3 5 4 8 8 9 9 2 4
5 6 8 8 2 0 3 3 6 1 9 3 8 6 4 7
9 9 2 6 7 8 7 5 2 5 9 1 1 5 3 5
6 7 0 1 8 4 1 2 9 7 5 1 2 6 2 5
8 1 5 4 4 0 0 8 6 6 6 5 2 5 3 5
1 5 0 4 4 5 0 6 5 8 7 3 2 1 0 0
3 1 3 0 0 8 0 3 8 0 6 4 6 9 7 9
7 1 7 7 0 2 0 0 4 1 3 5 4 5 9 4
4 5 3 6 6 8 1 7 6 0 3 8 7 2 5 8
2 1 0 1 9 0 0 9 2 4 9 4 7 3 3 2
9 1 8 5 0 3 4 5 4 6 5 6 0 4 9 4
3 4 1 0 3 6 6 9 6 2 9 8 3 7 5 5
9 4 7 2 1 7 4 3 5 7 7 8 4 9 2
0 9 5 7 9 9 3 2 6 3 2 3 6 1 7 0
0 7 5 8 3 1 7 0 3 4 9 5 2 1 8 8
0 3 9 6 8 4 9 1 7 6 5 6 9 9 0 2
```

- [ ] 597036
- [ ] 693646
- [ ] 367659
- [ ] 430713
- [ ] 807268
- [ ] 120457
- [ ] 527830
- [ ] 579932
- [ ] 115114
- [ ] 208405
- [ ] 473186
- [ ] 926564
- [ ] 751431
- [ ] 100001
- [ ] 962983
- [ ] 622218

Puzzle 64

```
2 7 8 3 7 6 1 0 9 3 0 9 0 6 0 7
9 7 7 4 8 5 3 5 9 8 1 2 3 9 3 6
3 5 8 0 6 1 8 9 2 6 2 1 2 1 4 5
4 6 1 7 2 0 8 3 3 2 6 1 3 5 1 4
1 3 0 1 1 4 8 6 6 4 0 7 6 1 6 4
9 5 0 9 5 0 2 1 6 1 4 4 7 5 8 1
2 2 8 0 7 0 3 3 2 2 1 7 2 6 6 6
8 2 1 3 9 2 7 2 9 4 8 7 6 3 2 7
3 9 6 9 1 2 5 6 9 2 0 6 9 1 9 4
5 7 4 3 9 9 6 7 2 1 1 4 0 0 7 0
4 9 6 4 4 9 8 1 3 9 3 7 7 4 8 2
1 4 8 1 4 4 7 0 4 1 3 7 8 1 2 7
2 8 1 5 2 0 6 4 1 4 2 6 3 7 0 2
5 3 5 0 6 8 6 4 9 9 0 3 7 6 4 0
5 1 1 9 0 3 1 5 8 3 2 6 8 9 4 9
4 9 7 7 1 9 8 1 8 3 1 6 9 2 5 2
9 5 4 3 5 6 7 3 5 0 9 8 3 8 0 7
0 5 1 0 7 0 1 2 0 1 3 8 8 7 9 6
```

- [ ] 971163
- [ ] 191493
- [ ] 060903
- [ ] 341509
- [ ] 748535
- [ ] 361326
- [ ] 837610
- [ ] 455214
- [ ] 832137
- [ ] 900144
- [ ] 804992
- [ ] 705115
- [ ] 833261
- [ ] 053765
- [ ] 310417
- [ ] 710329

Puzzle 65

```
8 1 3 9 6 9 2 5 2 8 0 0 0 8 1 7
3 6 9 6 5 3 0 4 8 8 0 5 6 6 4 7
9 2 1 5 6 6 8 4 5 5 7 4 5 8 4 9
6 3 5 2 6 6 9 3 3 8 6 2 5 6 2 5
6 4 0 8 1 2 3 2 7 6 0 4 9 7 4 5
2 3 6 1 1 1 1 2 6 2 8 7 0 9 7 8
4 6 9 0 8 9 1 1 0 7 2 2 5 6 7 2
0 9 6 9 1 8 4 5 7 6 5 5 2 9 6 9
4 8 9 6 1 9 3 1 4 6 5 3 3 6 4 0
8 7 4 6 7 6 2 3 8 2 6 2 4 5 4 3
3 0 3 4 7 9 1 8 1 8 5 2 5 5 3 0
1 9 0 9 1 3 9 6 2 0 1 3 2 1 4 9
1 6 7 5 3 2 8 2 9 9 3 9 3 6 0 7
7 0 5 2 9 7 3 1 8 8 6 7 7 3 8 8
8 3 9 7 9 0 7 7 1 2 0 0 7 0 0 8
1 0 4 3 2 6 2 7 2 2 0 3 9 0 5 9
5 8 6 7 3 5 2 5 3 7 3 1 5 6 0 7
8 7 6 5 8 7 4 2 8 2 3 9 1 5 3 2
```

☐ 926753	☐ 338625	☐ 272623	☐ 211154
☐ 673811	☐ 735253	☐ 666799	☐ 960519
☐ 048805	☐ 569697	☐ 557458	☐ 181166
☐ 307594	☐ 247764	☐ 692528	☐ 223970

Puzzle 66

```
4 9 8 9 0 0 7 0 6 3 9 0 5 8 6 6
8 2 0 2 1 2 2 6 8 1 8 7 8 5 2 5
0 6 4 0 6 5 8 6 7 9 4 3 1 8 9 0
6 1 7 0 3 0 7 0 5 7 9 9 3 0 0 0
0 3 5 0 3 0 6 0 2 1 2 1 5 2 4 2
9 9 9 3 0 9 1 2 8 3 2 2 1 2 7 4
6 7 7 8 7 9 2 1 2 7 6 8 7 5 5 3
1 7 3 0 1 2 1 4 7 3 6 3 5 9 0 3
4 3 0 4 4 3 4 7 8 8 9 8 7 4 8 6
9 9 2 9 5 9 9 7 9 1 3 6 8 7 4 2
3 7 6 3 5 9 3 0 8 5 0 8 7 7 4 7
7 0 9 7 1 1 9 2 8 1 5 5 8 1 5 2
8 9 6 5 9 4 0 0 2 9 7 0 8 9 3 2
3 7 1 5 5 0 7 0 6 7 3 6 3 8 1 9
5 3 9 0 0 2 9 3 2 8 1 0 0 6 5 0
7 9 6 3 2 1 1 7 6 3 0 7 6 3 3 5
4 6 5 1 5 9 7 2 8 9 6 7 1 8 8 0
8 4 4 3 5 9 5 0 8 6 1 3 1 5 4 7
```

☐ 215620	☐ 795503	☐ 280236	☐ 153724
☐ 622396	☐ 914021	☐ 830671	☐ 301178
☐ 394079	☐ 878757	☐ 508445	☐ 289671
☐ 070098	☐ 696196	☐ 709739	☐ 818785

Puzzle 67

```
5 9 8 5 7 3 2 7 8 2 4 7 5 4 9 1
7 1 2 5 8 9 6 3 6 1 6 7 5 3 5 2
1 4 6 3 2 2 5 3 3 4 5 4 3 2 6 1
5 2 2 3 2 4 3 8 7 5 1 9 8 9 1 1
4 8 8 3 6 6 6 5 4 3 6 3 5 4 8 8
7 5 4 0 4 2 8 3 8 0 4 3 7 8 5 9
1 9 5 7 6 7 7 5 5 7 5 3 5 2 0 0
4 3 6 7 1 9 2 8 1 5 5 2 4 5 0 0
4 5 6 7 8 3 8 7 7 6 9 4 8 0 4 2
0 2 8 1 3 4 8 0 2 2 9 3 0 5 0 1
2 5 5 8 8 8 6 7 7 7 1 4 6 6 5 2
0 8 6 0 4 8 4 8 5 8 4 2 5 9 5 9
3 2 7 6 2 0 9 6 6 9 2 8 9 0 0 2
4 2 7 4 0 1 5 6 9 2 0 6 9 0 3 6
9 2 7 1 1 5 4 2 9 9 3 1 1 6 8 1
2 5 6 8 0 4 3 3 2 6 5 0 1 5 9 7
9 6 0 5 8 3 5 6 2 0 4 7 8 7 3 9
1 6 8 7 0 8 0 0 0 3 8 7 5 7 5 1
```

☐ 891578	☐ 793488	☐ 155245	☐ 025357
☐ 232685	☐ 727569	☐ 302044	☐ 574287
☐ 253958	☐ 963616	☐ 276209	☐ 274015
☐ 154299	☐ 659911	☐ 146322	☐ 727432

Puzzle 68

```
6 7 5 3 5 1 4 3 2 4 0 6 4 9 6 5
6 3 3 9 0 6 4 1 9 7 9 1 3 6 0 1
3 7 1 3 4 6 5 2 5 9 9 0 9 7 9 8
2 0 7 4 5 3 6 1 4 8 6 8 8 7 3 5
6 8 7 9 4 3 1 8 8 8 0 2 7 9 1 9
3 5 8 1 5 8 4 4 8 5 1 8 4 4 3 2
1 8 0 4 6 1 6 1 9 5 9 6 4 5 4 0
7 3 1 5 0 0 7 1 5 5 7 1 7 7 2 4
4 8 3 7 6 1 7 5 9 1 0 2 6 0 1 6
1 4 1 3 6 0 8 7 7 3 9 7 7 8 1 6
1 3 2 1 7 0 9 1 6 4 5 7 2 6 7 7
2 8 0 7 9 8 9 2 3 4 6 5 7 5 2 2
9 3 3 9 0 3 7 2 5 9 4 6 2 7 1 1
5 1 0 5 7 9 7 6 0 8 1 9 8 2 9 3
1 4 5 9 5 7 7 6 0 4 5 8 3 9 9 6
6 8 8 4 3 8 4 2 4 4 3 8 8 0 8 6
1 7 4 5 0 2 3 8 1 2 4 0 1 5 3 0
3 6 9 6 8 7 8 0 1 3 1 4 1 9 1 8
```

☐ 088965	☐ 158084	☐ 495273	☐ 093134
☐ 850302	☐ 545606	☐ 795988	☐ 419439
☐ 136087	☐ 858073	☐ 218411	☐ 542699
☐ 579760	☐ 406775	☐ 643298	☐ 663381

Puzzle 69

```
5 5 3 7 5 8 2 0 3 1 2 1 2 8 2 7
1 4 6 2 2 0 4 2 6 9 9 0 5 8 1 4
2 4 3 8 1 1 8 2 3 7 5 4 7 3 2 1
8 2 8 0 5 9 6 6 5 2 0 8 5 3 4 5
6 1 9 7 0 0 3 2 1 2 2 6 5 2 3 1
0 2 4 8 2 9 8 8 2 4 8 4 2 0 5 5
9 2 9 4 4 8 2 5 8 8 4 0 7 5 8 3
9 1 1 1 6 4 1 7 3 5 4 3 2 8 5 6
6 2 9 7 1 0 8 1 2 8 3 4 8 2 4 4
1 5 7 3 4 1 6 6 1 1 3 3 5 7 8 6
0 3 0 2 8 3 0 1 1 2 1 5 5 1 3 0
4 8 5 9 6 9 5 5 4 8 4 9 3 4 9 6
2 6 8 2 3 9 8 2 8 6 3 1 1 2 8 8
5 6 2 3 3 2 6 6 1 8 9 0 0 6 0 1
6 2 2 5 2 1 9 2 3 2 8 6 9 9 6 0
1 6 0 9 6 3 8 9 4 5 8 7 9 8 0 4
4 9 3 5 3 2 4 1 2 6 4 7 5 7 5 5
4 9 0 8 6 9 5 3 9 5 7 9 3 5 4 1
```

☐ 239828	☐ 614696	☐ 870827	☐ 534212
☐ 970582	☐ 058593	☐ 582714	☐ 174352
☐ 211182	☐ 508614	☐ 111926	☐ 638945
☐ 344820	☐ 383532	☐ 816844	☐ 221244

Puzzle 70

```
9 5 0 1 3 0 2 5 4 4 7 4 6 4 1 7
3 3 7 2 0 5 7 6 5 6 1 5 9 1 5 5
3 6 7 8 8 8 2 9 2 8 7 6 7 3 2 5
4 1 5 3 1 0 6 1 2 8 5 5 0 0 6 8
2 2 3 8 3 0 7 9 0 5 6 4 7 1 6 5
6 6 7 4 0 8 6 6 2 8 2 0 8 0 8 7
3 3 3 9 9 2 6 1 0 3 0 3 8 9 5 9
7 5 7 4 4 7 7 0 3 3 1 1 4 7 2 8
9 1 1 8 3 8 1 1 5 2 3 4 0 5 3 1
0 9 7 7 2 5 6 9 2 7 9 6 4 9 2 0
3 0 0 6 4 6 4 7 8 6 1 4 2 9 8 8
3 9 9 6 5 3 3 3 3 0 3 8 9 9 7 7
3 7 7 7 1 8 6 0 1 1 6 4 1 6 5 6
7 7 0 2 3 3 4 1 9 1 1 3 9 4 5 6
3 4 3 3 8 4 1 9 4 6 8 9 1 2 7 9
0 1 2 8 0 0 5 6 7 0 4 0 6 3 9 2
7 2 3 6 4 7 8 1 3 7 8 2 1 6 8 0
3 7 3 1 2 7 6 0 8 7 9 7 8 5 4 8
```

☐ 461766	☐ 327601	☐ 429191	☐ 975999
☐ 765615	☐ 361316	☐ 678494	☐ 485671
☐ 841946	☐ 744757	☐ 735770	☐ 668523
☐ 103038	☐ 348464	☐ 147790	☐ 362163

Puzzle 71

```
3 9 5 4 8 1 0 3 7 0 4 4 5 2 6 3
7 6 8 5 0 4 0 0 7 9 1 4 0 2 7 0
0 5 6 8 0 3 5 6 3 1 6 4 2 1 7 2
5 5 3 5 9 6 5 3 4 5 8 1 9 9 5 8
6 9 0 8 5 8 1 8 4 3 5 1 8 7 0 4
0 9 9 7 0 2 2 4 6 2 1 6 5 3 5 4
4 3 4 3 4 0 0 4 6 3 2 9 7 5 2 1
2 3 7 3 0 8 7 5 5 3 5 2 3 7 6 8
0 9 4 0 6 3 3 2 6 5 6 3 7 1 6 4
1 5 2 7 7 7 3 7 5 7 4 1 4 5 7
9 3 3 7 5 7 1 3 3 8 0 5 2 2 4 7
5 6 7 3 1 3 2 7 9 8 3 0 3 3 0 5
6 8 9 2 0 5 6 9 9 1 4 6 6 6 0 6
1 1 4 2 5 5 7 0 9 5 9 2 7 2 0 1
4 6 7 3 7 5 5 3 2 4 6 3 5 0 5 2
3 9 8 4 1 0 8 1 4 5 2 7 8 2 8 8
4 1 3 5 4 6 9 4 0 3 9 8 1 7 4 4
7 9 2 4 5 5 0 2 0 1 3 5 0 8 7 5
```

☐ 504007	☐ 316421	☐ 074833	☐ 599185
☐ 650062	☐ 777251	☐ 533055	☐ 370560
☐ 923640	☐ 843518	☐ 800727	☐ 254407
☐ 783919	☐ 571423	☐ 425248	☐ 245575

Puzzle 72

```
4 2 0 1 0 3 2 5 7 1 1 4 6 0 2 6
7 6 6 8 0 6 9 0 8 3 3 0 2 9 7 0
9 3 9 7 0 7 8 8 3 2 3 2 2 3 1 4
1 0 9 1 9 8 1 4 6 0 1 2 8 1 1 0
2 1 7 7 6 0 4 2 4 7 8 8 1 3 5 8
3 8 0 7 4 1 0 4 4 4 6 6 4 5 6 8
8 3 1 8 5 3 9 3 2 1 2 6 4 0 2 0
1 5 7 6 9 0 7 8 9 9 5 3 2 2 2 9
0 7 9 2 6 4 2 9 7 5 5 2 9 6 4 1
6 5 2 2 6 6 9 2 0 8 2 6 7 1 7 6
8 7 9 4 4 7 9 8 5 9 9 9 5 2 0 9
0 9 6 4 3 7 1 0 1 6 2 8 3 7 4 0
3 4 2 5 6 6 8 1 2 1 4 3 4 7 6 9
6 2 6 9 1 7 6 2 7 7 2 1 4 9 0 3
9 6 5 3 1 8 2 1 2 0 9 3 6 0 5 9
3 1 5 4 6 3 6 7 8 4 4 4 7 3 7 4
9 9 8 2 7 4 6 5 3 1 5 4 9 8 9 6
3 8 4 3 2 4 5 4 1 0 7 9 6 9 9 4
```

☐ 463678	☐ 757538	☐ 427251	☐ 069083
☐ 397437	☐ 479859	☐ 842438	☐ 564728
☐ 671962	☐ 053139	☐ 261017	☐ 295657
☐ 668121	☐ 792462	☐ 622814	☐ 742265

Puzzle 73

```
5 8 8 2 8 7 6 5 4 0 9 9 1 6 6 4
2 2 0 7 7 7 0 0 6 9 3 2 8 3 9 9
3 1 7 3 6 9 0 3 1 4 7 3 2 0 4 4
3 5 2 5 6 5 4 3 1 5 9 5 5 3 6 8
6 7 3 4 4 6 3 8 1 0 2 2 4 9 2 5
6 6 7 7 4 2 7 7 4 8 1 9 2 6 5 0
2 6 6 2 7 4 2 1 8 1 8 4 0 5 5 7
0 7 4 6 6 7 5 9 8 5 0 7 7 4 8 0
0 9 5 5 7 1 6 2 4 7 2 7 7 5 2 6
9 1 7 2 6 9 4 4 3 4 2 7 4 6 9 6
9 8 3 7 0 8 4 1 6 1 0 3 9 7 3 1
6 0 3 5 7 8 6 4 9 1 2 7 8 7 9 8
5 4 8 0 3 0 4 6 2 2 9 5 1 3 8 4
6 9 1 8 5 3 2 9 6 6 2 7 7 9 2 4
4 5 0 8 0 2 6 3 7 8 8 2 6 9 0 5
2 0 8 6 2 8 1 8 9 3 5 0 5 4 8 3
2 1 7 6 9 6 9 2 3 4 5 8 6 2 6 4
6 9 9 6 6 4 6 7 9 2 7 7 0 8 8 8
```

- ☐ 725644
- ☐ 764573
- ☐ 866686
- ☐ 855264
- ☐ 220183
- ☐ 947702
- ☐ 354726
- ☐ 461976
- ☐ 770069
- ☐ 545677
- ☐ 900266
- ☐ 192225
- ☐ 031473
- ☐ 671872
- ☐ 182680
- ☐ 176969

Puzzle 74

```
8 6 5 1 1 2 2 2 7 2 3 0 6 8 2 9
9 0 1 6 5 9 4 0 4 5 6 5 3 5 5 9
6 9 6 1 5 4 7 0 8 9 4 8 6 1 3 2
7 8 6 4 2 1 8 1 2 3 7 6 8 6 5 1
8 1 7 4 0 2 9 0 4 0 4 8 8 6 4 7
5 0 4 4 5 4 0 8 2 0 9 9 3 2 3 6
0 5 4 7 9 1 1 1 6 9 3 6 3 9 2 6
5 7 3 0 4 0 3 8 2 1 4 1 4 0 6 3
9 9 5 6 5 4 5 2 0 2 7 8 4 3 8 6
3 9 4 3 8 5 5 9 4 4 1 9 8 0 0 8
3 7 5 6 7 7 1 7 7 7 3 3 5 5 8 7
8 2 7 9 6 3 5 6 8 1 3 2 0 3 3 4
9 9 3 5 4 3 6 7 0 9 4 0 2 6 3 8
5 0 4 5 8 1 4 9 9 2 7 2 8 9 5 1
5 7 0 3 2 9 8 2 8 6 8 3 9 5 3 1
8 2 2 4 4 2 9 5 7 3 5 2 1 8 2 8
4 8 1 9 0 1 1 9 4 7 5 2 1 9 0 3
5 8 1 1 6 2 9 6 4 0 7 5 0 0 4 2
```

- ☐ 242620
- ☐ 181246
- ☐ 445408
- ☐ 208349
- ☐ 892417
- ☐ 798754
- ☐ 304037
- ☐ 537369
- ☐ 149599
- ☐ 547911
- ☐ 588492
- ☐ 553565
- ☐ 808623
- ☐ 335971
- ☐ 764824
- ☐ 896785

Puzzle 75

```
2 1 4 6 9 0 7 5 4 2 3 7 2 4 2 1
8 2 3 5 8 2 5 1 4 5 6 0 9 4 6 8
1 2 2 8 7 6 4 6 2 7 4 6 4 2 0 7
3 3 9 2 0 6 8 6 1 6 5 4 5 9 5 4
0 8 5 6 0 1 0 0 9 6 0 2 6 9 6 7
7 7 4 5 3 4 1 4 5 9 1 5 2 0 2 5
0 7 7 0 2 2 4 1 4 0 7 8 8 0 5 7
1 5 2 5 7 3 1 4 4 5 0 9 9 9 1 5
8 2 8 3 8 1 1 6 6 2 0 4 3 3 3 0
7 2 5 7 3 5 9 5 7 4 1 4 8 0 0 3
0 5 5 7 3 6 9 4 9 2 3 5 1 7 0 7
6 7 7 0 5 0 7 1 2 9 5 6 1 8 8 2
1 6 0 7 6 7 5 5 4 0 6 9 4 9 6 1
0 2 2 2 6 7 2 3 5 3 9 0 3 5 1 0
0 9 8 5 0 9 5 4 9 1 5 4 8 9 4 2
1 5 6 0 3 7 9 5 2 0 1 4 0 4 6 0
5 4 9 3 7 4 4 0 9 7 3 6 7 6 1 1
7 9 0 7 8 7 5 9 8 9 8 4 9 3 6 8
```

- ☐ 415345
- ☐ 735957
- ☐ 427324
- ☐ 083411
- ☐ 285328
- ☐ 444022
- ☐ 827459
- ☐ 644591
- ☐ 682075
- ☐ 607810
- ☐ 126058
- ☐ 999054
- ☐ 440965
- ☐ 429900
- ☐ 963216
- ☐ 469075

Puzzle 76

```
1 1 2 1 0 6 4 8 0 1 1 4 0 7 5 4
3 9 0 5 4 6 2 0 2 3 1 3 6 6 7 6
6 3 0 6 3 5 1 4 5 3 5 5 2 7 6 8
0 7 7 7 6 5 4 7 7 9 7 8 2 8 8
6 0 1 1 3 5 8 0 1 9 6 3 5 2 1 1
4 5 6 2 6 3 4 4 2 2 5 4 8 8 5 9
1 2 3 3 1 9 5 4 3 4 9 1 1 6 5 1
0 3 7 3 8 0 4 2 6 4 0 9 4 9 3 6
3 7 0 8 5 4 7 4 3 1 9 1 4 7 4 4
1 7 3 7 3 0 7 3 4 4 2 8 0 2 1 8
9 4 8 4 4 4 6 6 5 8 8 0 6 2 8 2
1 6 4 7 0 9 9 1 7 7 0 6 2 9 1 9
2 8 2 4 0 0 9 8 5 3 0 0 8 2 4 0
9 6 4 4 5 2 0 6 1 7 7 4 1 5 3 6
3 6 2 8 7 3 6 1 3 4 0 3 4 7 1 1
6 5 6 3 5 3 6 7 8 2 9 8 4 8 2 9
4 4 2 0 2 7 4 6 0 8 5 6 4 5 7 8
6 1 4 1 5 4 5 1 4 9 8 6 9 7 9 5
```

- ☐ 720244
- ☐ 445760
- ☐ 961733
- ☐ 437037
- ☐ 673734
- ☐ 570448
- ☐ 689434
- ☐ 044185
- ☐ 191293
- ☐ 085474
- ☐ 842426
- ☐ 253370
- ☐ 181431
- ☐ 846012
- ☐ 551867
- ☐ 429733

Puzzle 77

```
3 3 5 2 6 0 6 7 0 0 5 7 4 8 9 1
0 8 5 5 5 6 7 1 5 5 5 0 7 8 8 8
8 3 7 0 2 6 5 0 6 6 3 9 4 9 9 5
5 5 0 6 9 3 5 1 8 0 0 3 0 4 1 7
3 9 9 4 6 7 0 1 5 0 1 7 5 8 7 0
2 9 1 7 9 1 1 0 3 9 1 8 4 0 1 0
8 4 4 7 6 2 0 1 2 6 6 3 5 0 8 4
2 0 2 9 5 4 2 0 2 7 7 2 3 8 7 5
2 1 4 9 5 8 1 0 2 7 1 5 4 8 7 4
1 5 2 6 7 3 3 6 3 9 3 6 4 6 6 3
6 1 0 2 5 2 4 9 4 1 2 2 8 7 2 4
8 6 0 1 0 4 2 7 2 3 3 7 6 8 2 0
0 5 8 0 9 5 5 5 5 2 2 5 7 4 1 7
6 0 5 4 2 1 0 4 6 3 9 8 9 1 5 1
1 9 4 1 0 8 8 9 9 8 0 0 1 5 3 5
5 5 6 8 7 1 9 6 0 8 9 8 7 0 4 7
6 9 6 1 1 0 0 5 3 9 0 3 0 6 0 6
2 8 7 4 3 8 6 3 9 7 7 2 3 6 0 5
```

☐ 965243 ☐ 419075 ☐ 090923 ☐ 590561
☐ 214958 ☐ 606253 ☐ 008867 ☐ 853222
☐ 197684 ☐ 301167 ☐ 883231 ☐ 010551
☐ 006975 ☐ 888705 ☐ 769006 ☐ 110053

Puzzle 78

```
8 0 0 5 9 9 2 7 5 1 6 5 9 8 0 9
2 3 8 8 4 2 6 6 3 5 1 7 9 0 1 2
4 7 6 3 1 6 4 7 0 9 6 2 9 0 6 9
8 3 4 5 7 8 8 3 6 2 5 1 7 3 7 5
1 4 6 2 6 6 2 7 9 0 5 6 5 2 1 7
7 5 9 5 0 4 2 4 1 5 2 3 3 3 8 0
0 6 5 1 2 8 1 9 3 2 8 0 1 8 7 8
0 7 5 0 5 1 7 8 8 1 4 3 1 6 1 8
5 1 4 6 2 8 1 9 3 3 0 5 6 8 9 2
4 2 1 6 5 1 2 9 2 3 0 8 9 9 1 4
4 9 2 9 6 1 0 5 1 6 5 0 6 2 8 3
9 9 1 1 2 3 6 5 4 1 5 8 6 3 4 0
6 5 1 2 0 6 0 2 6 5 7 1 6 6 6 9
8 4 3 2 2 7 4 6 5 9 2 6 3 2 0 2
7 4 9 5 4 2 3 2 8 6 1 4 8 3 4 2
5 2 3 4 0 1 1 1 5 1 3 7 9 5 3 6
7 3 1 1 9 4 6 4 6 2 3 5 4 4 0 0
5 7 7 3 9 2 3 3 5 2 0 8 9 4 5 6
```

☐ 734567 ☐ 142194 ☐ 788362 ☐ 123831
☐ 676020 ☐ 248832 ☐ 256202 ☐ 865033
☐ 926425 ☐ 735165 ☐ 406481 ☐ 689236
☐ 316470 ☐ 005992 ☐ 921561 ☐ 746592

Puzzle 79

```
7 3 3 0 7 6 0 1 8 3 2 0 1 9 7 4
4 7 2 5 6 8 0 0 8 7 8 1 0 8 9 2
0 1 6 6 0 9 5 9 7 2 3 1 5 3 8 4
4 9 3 6 4 1 3 9 8 5 9 1 5 7 0 5
1 9 6 0 7 8 2 5 1 4 2 7 0 9 7 4
5 4 4 3 5 2 6 6 1 1 1 2 3 4 5 8
5 0 7 4 0 2 5 3 2 2 3 7 0 1 3 2
3 4 5 3 9 6 6 1 3 6 1 5 0 5 0 7
7 4 7 2 9 7 9 4 1 0 2 4 0 1 8 9
3 2 6 0 0 2 0 5 6 4 6 3 9 0 6 7
4 6 0 2 9 2 1 9 0 8 8 4 7 7 9 8
4 2 5 3 9 2 1 5 2 1 6 3 1 5 1 8
1 1 7 6 1 3 3 3 3 0 2 9 3 7 0 2
4 3 6 7 8 0 7 1 6 3 0 9 8 0 7 1
9 9 0 4 5 3 5 3 5 5 7 0 3 0 4 9
4 0 5 7 0 8 4 4 4 4 4 1 2 5 9 8
1 0 5 2 3 1 8 7 5 9 0 4 7 7 4 8
8 0 1 1 1 5 5 4 4 0 5 7 3 8 7 6
```

☐ 739215 ☐ 499173 ☐ 480750 ☐ 411527
☐ 477409 ☐ 293774 ☐ 780086 ☐ 092315
☐ 330293 ☐ 326262 ☐ 131293 ☐ 018320
☐ 505202 ☐ 070196 ☐ 748809 ☐ 510757

Puzzle 80

```
0 1 4 0 0 3 2 3 7 3 0 7 2 9 3 0
9 0 4 2 7 8 8 4 0 9 9 6 3 4 0 2
8 3 3 8 2 7 1 2 8 8 4 3 3 0 5 5
9 6 1 2 4 3 1 4 8 6 5 7 4 0 5 2
1 7 6 7 1 7 6 2 0 1 2 4 7 3 6 4
9 1 7 9 5 7 5 8 2 6 6 4 1 4 0 6
7 4 7 1 7 5 7 2 1 2 6 5 1 2 7 8
2 9 6 4 1 8 2 3 5 8 6 2 9 3 3 1
8 3 1 8 3 7 8 2 6 8 1 4 4 0 3 2
0 0 0 9 0 6 3 5 9 4 8 7 6 4 0 0
1 9 9 1 4 6 0 3 3 8 1 0 9 0 5 2
9 8 0 8 0 4 6 6 2 4 8 4 9 9 0 5
1 3 0 6 1 1 3 7 8 9 3 3 1 4 1 6
8 1 1 9 0 1 5 7 4 0 3 9 9 1 2 5
5 1 4 4 5 0 1 2 3 7 9 5 5 5 8 9
4 4 1 3 9 2 2 8 3 6 3 0 1 7 3 0
0 9 1 2 3 3 9 0 0 5 0 1 1 1 1 4
4 0 9 7 3 1 9 5 7 4 8 7 6 5 7 4
```

☐ 950104 ☐ 012379 ☐ 271288 ☐ 117332
☐ 893314 ☐ 900141 ☐ 287381 ☐ 191854
☐ 523282 ☐ 442660 ☐ 005011 ☐ 501283
☐ 417630 ☐ 072415 ☐ 904887 ☐ 034393

Puzzle 81

```
8 3 3 6 9 3 5 7 8 7 8 4 3 0 6 8
1 7 7 4 3 9 9 1 8 0 8 6 5 1 9 6
9 4 0 0 4 2 4 9 7 4 2 6 5 1 8 8
0 9 4 7 1 5 6 2 7 3 3 9 9 0 7 5
0 9 0 4 0 9 0 5 8 3 1 8 1 2 6 5
3 3 9 7 9 8 8 1 8 1 7 5 2 1 7 7
9 9 5 0 5 0 8 6 5 9 1 2 0 3 2 5
0 9 1 6 0 8 9 8 9 0 2 7 7 5 7 6
8 0 3 4 4 2 0 6 3 9 6 8 9 3 1 3
5 4 8 9 0 8 0 9 3 9 0 5 3 1 6 6
0 0 9 2 3 2 3 6 3 0 4 1 8 1 0 6
0 5 9 7 1 0 1 4 6 8 3 9 5 1 5 6
6 8 7 3 3 6 2 6 6 9 2 4 3 5 1 8
0 6 7 2 3 4 2 2 2 0 0 7 8 9 3 2
5 3 6 6 9 3 1 6 3 7 2 6 9 8 0 9
3 9 9 5 9 6 6 9 8 1 1 8 2 6 8 7
4 1 6 1 9 8 1 8 6 7 9 4 2 9 3 3
0 1 8 0 8 8 3 8 1 1 7 8 6 1 4 8
```

- [ ] 683266
- [ ] 050599
- [ ] 199347
- [ ] 881817
- [ ] 276789
- [ ] 987002
- [ ] 118560
- [ ] 583181
- [ ] 839493
- [ ] 821626
- [ ] 878430
- [ ] 699015
- [ ] 808651
- [ ] 430369
- [ ] 265174
- [ ] 809382

Puzzle 82

```
2 0 0 3 1 8 4 0 1 1 1 0 1 2 2 0
0 8 9 3 5 6 1 6 1 9 4 9 6 0 9 9
2 4 0 5 5 6 1 0 8 5 8 1 5 3 1 0
7 2 0 0 9 4 0 8 9 6 5 7 7 6 4 6
6 2 9 5 2 3 9 3 0 2 1 2 9 2 0 7
1 0 1 3 0 9 4 2 1 3 3 4 9 4 8 0
8 1 2 6 4 0 5 9 5 9 5 2 7 2 8 7
8 4 5 8 4 4 1 3 7 7 7 0 4 4 8 8
0 1 5 4 0 7 3 9 9 1 0 4 7 2 0 3
2 6 8 0 5 4 1 3 2 4 8 1 7 0 8 7
3 0 2 8 8 3 4 5 9 4 1 0 7 9 3 9
3 2 2 6 6 4 6 8 8 3 1 5 5 7 3 6
3 5 4 1 1 7 2 0 1 9 4 4 4 6 1 1
2 3 9 1 1 9 3 8 3 7 3 9 7 1 5 1
8 7 2 7 2 3 7 3 0 7 6 8 7 6 3 2
7 1 2 6 6 6 6 0 5 9 9 4 6 8 3 1
4 8 2 4 2 4 1 7 8 6 4 6 8 9 8 1
3 7 6 0 1 4 0 9 5 1 9 5 8 1 2 9
```

- [ ] 349718
- [ ] 635053
- [ ] 091255
- [ ] 302883
- [ ] 463974
- [ ] 580165
- [ ] 916165
- [ ] 203624
- [ ] 864689
- [ ] 302129
- [ ] 808880
- [ ] 861679
- [ ] 394154
- [ ] 001924
- [ ] 823332
- [ ] 680541

Puzzle 83

```
2 0 1 9 5 3 9 0 4 7 7 9 9 4 6 6
2 8 5 8 9 2 3 1 1 0 7 5 1 9 1 5
1 2 2 1 7 6 0 4 4 3 7 5 7 5 5 5
6 4 9 3 1 8 9 3 0 5 1 4 4 2 7 2
0 3 6 0 4 0 6 8 8 1 1 3 8 4 5 2
6 3 3 4 4 5 6 8 8 8 7 0 9 4 2 9
4 3 0 9 3 6 8 4 0 8 1 4 6 7 9 4
8 2 1 2 5 5 5 4 4 0 6 5 2 1 6 1
6 8 2 3 3 5 0 3 6 6 6 7 5 8 6 7
9 6 1 0 4 9 0 1 8 8 0 9 0 2 2 7
8 2 7 6 0 8 6 3 5 7 3 7 5 6 4 4
9 6 9 1 1 6 6 5 5 6 7 0 3 3 9 8
1 4 6 5 3 9 2 0 7 7 3 9 7 3 0 8
3 1 0 0 1 0 5 8 9 1 9 4 7 3 2 5
9 5 8 9 9 2 3 0 7 2 0 5 9 3 8 6
4 0 9 4 7 0 8 1 7 4 3 8 0 4 0 9
8 1 0 7 3 7 2 0 1 1 6 7 8 3 0 2
9 6 9 3 7 3 9 7 8 5 0 7 9 3 2 9
```

- [ ] 090668
- [ ] 049018
- [ ] 923110
- [ ] 881138
- [ ] 125604
- [ ] 596008
- [ ] 093591
- [ ] 630121
- [ ] 249078
- [ ] 375368
- [ ] 705916
- [ ] 490280
- [ ] 606486
- [ ] 743804
- [ ] 406712
- [ ] 773973

Puzzle 84

```
7 8 3 4 9 7 3 2 0 5 7 2 9 8 0 7
7 3 4 9 1 8 9 6 1 7 6 6 1 4 4 4
9 2 0 4 6 5 1 3 2 5 8 2 1 1 7 6
2 0 3 6 8 2 8 7 4 4 7 2 5 0 4 5
6 4 1 3 1 6 1 5 0 9 2 0 2 9 8 7
9 9 9 2 6 9 7 9 4 1 2 6 2 0 6 4
0 7 5 1 3 1 6 1 2 3 2 2 5 2 7 5
6 3 7 7 9 5 2 9 1 2 5 0 5 9 5 1
6 5 4 5 8 6 9 6 8 0 0 6 5 1 8 3
2 7 4 4 8 6 9 0 6 9 1 5 1 8 5 9
1 6 1 6 8 2 1 2 0 3 4 4 7 6 9 6
4 0 9 9 9 4 9 6 3 0 5 4 6 6 6 9
3 3 9 1 3 3 6 3 1 3 5 6 3 1 0 1
5 1 1 3 9 8 4 5 6 6 9 0 3 7 8 1
6 7 9 7 2 2 4 8 6 9 4 6 9 5 8 6
0 8 8 9 0 4 8 9 3 7 4 5 0 3 2 7
6 9 8 3 1 2 4 4 1 8 9 5 7 4 3 9
1 3 4 5 8 8 4 5 6 2 2 1 4 8 2 2
```

- [ ] 181762
- [ ] 984098
- [ ] 766648
- [ ] 719384
- [ ] 645712
- [ ] 906621
- [ ] 768474
- [ ] 923396
- [ ] 829359
- [ ] 541052
- [ ] 918661
- [ ] 616875
- [ ] 262149
- [ ] 918961
- [ ] 691928
- [ ] 202905

Puzzle 85

```
2 9 4 1 1 9 1 5 8 5 0 2 6 7 7 5
4 6 1 9 5 3 2 0 5 0 7 5 3 2 6 8
6 1 3 6 6 4 2 5 1 5 3 8 7 2 8 6
2 4 7 6 1 2 0 5 1 9 5 9 7 6 6 5
3 0 7 8 6 7 8 6 9 8 1 6 5 3 4 5
1 9 3 8 1 7 1 9 8 7 1 1 0 7 6 8
2 1 6 6 7 2 5 4 4 3 6 0 1 3 3 2
5 4 0 3 0 8 8 7 0 2 3 5 8 5 2 1
0 5 5 8 1 7 8 6 7 7 6 8 1 9 2 4
8 4 1 0 6 7 5 8 7 5 4 1 3 7 8 8
0 1 9 1 5 3 9 9 8 4 8 4 3 5 5 2
8 4 4 0 7 5 7 7 0 8 5 2 8 5 0 2
1 7 8 9 6 1 3 3 4 8 1 0 1 1 8 4
9 1 7 3 0 9 2 9 0 4 0 0 8 0 7 5
3 8 9 4 7 3 4 7 0 5 8 9 8 3 4 7
2 2 7 5 8 2 6 1 9 5 0 1 8 2 3 6
7 6 5 9 5 3 6 7 6 9 0 5 8 3 8 0
6 5 8 6 3 7 6 5 3 5 3 7 8 0 6 4
```

- [ ] 101911
- [ ] 704891
- [ ] 515246
- [ ] 368011
- [ ] 789613
- [ ] 101843
- [ ] 502167
- [ ] 166725
- [ ] 850967
- [ ] 875413
- [ ] 903719
- [ ] 171839
- [ ] 235916
- [ ] 810591
- [ ] 708004
- [ ] 915399

Puzzle 86

```
4 2 2 4 9 7 5 9 7 1 3 5 4 5 9 3
8 9 6 8 3 5 4 9 9 6 3 9 9 3 0 4
0 3 8 4 1 4 2 6 1 0 1 2 2 7 9 3
0 3 4 4 7 9 6 2 6 6 0 2 2 7 0 5
7 2 4 1 4 2 5 0 9 4 3 3 1 4 7 8
2 2 0 3 4 6 5 3 8 7 3 0 0 0 4 2
6 8 9 9 9 5 4 9 2 8 3 9 2 3 8 9
8 9 0 4 9 7 9 9 1 8 0 7 2 8 7 9
0 2 7 2 4 8 8 3 2 6 9 8 2 0 6 3
5 7 4 0 0 0 7 7 3 4 4 4 5 7 1 1
9 7 4 3 1 3 9 7 2 4 2 8 5 2 3 3
8 4 6 2 4 4 4 4 3 8 2 8 7 7 7 7
9 1 1 6 0 5 7 7 4 9 8 5 3 4 4 4
1 2 3 6 1 4 4 5 9 6 2 5 6 3 0 3
0 7 2 0 8 0 8 2 1 7 6 8 0 4 5 0
3 6 7 6 5 3 8 6 1 1 5 8 9 2 1 8
6 6 0 9 8 3 4 5 0 8 0 4 8 0 1 4
8 0 7 4 9 7 1 4 3 6 8 0 1 5 3 2
```

- [ ] 739930
- [ ] 269544
- [ ] 313972
- [ ] 345080
- [ ] 990030
- [ ] 611589
- [ ] 644894
- [ ] 916487
- [ ] 808217
- [ ] 737922
- [ ] 137405
- [ ] 960662
- [ ] 927741
- [ ] 044927
- [ ] 280203
- [ ] 341459

Puzzle 87

```
7 2 1 5 0 4 7 9 3 1 3 9 0 6 4 3
8 3 3 7 8 0 2 8 6 1 1 3 5 5 9 4
7 2 6 4 2 2 1 7 2 8 5 9 3 3 6 1
7 4 1 4 6 8 8 9 1 6 1 1 1 1 1 3
5 4 7 3 7 3 9 5 7 8 2 4 8 4 8 2
1 6 5 4 6 3 2 4 5 0 7 5 0 9 1 5
0 9 1 4 9 9 4 5 6 9 3 2 0 6 1 8
9 8 2 9 3 9 2 8 8 7 5 8 9 6 6 1
1 5 5 2 2 7 6 1 3 0 6 6 6 9 1 2
9 8 8 6 5 2 4 1 6 4 0 3 8 7 3 4
9 6 8 3 8 1 7 5 5 2 7 3 1 0 4 0
8 9 9 8 5 9 3 8 5 7 7 7 7 3 6 4
2 4 0 9 8 9 6 3 7 3 5 3 1 5 6 0
0 5 0 2 2 7 8 5 0 4 1 7 6 2 6 9
0 8 1 8 9 1 6 1 2 7 5 3 8 7 7 8
0 1 1 0 2 1 7 3 4 9 1 4 3 2 1 8
2 2 2 1 9 0 3 1 3 8 2 5 4 4 0 9
6 7 6 2 1 3 4 1 0 8 8 0 5 9 7 9
```

- [ ] 452831
- [ ] 877510
- [ ] 028611
- [ ] 678888
- [ ] 372612
- [ ] 149669
- [ ] 371201
- [ ] 054236
- [ ] 091222
- [ ] 158351
- [ ] 697035
- [ ] 737395
- [ ] 577065
- [ ] 446985
- [ ] 997219
- [ ] 298362

Puzzle 88

```
8 6 2 3 6 7 6 1 3 1 2 1 4 1 4 2
9 6 1 9 2 7 8 3 7 7 6 0 5 4 1 4
3 2 9 9 9 2 6 6 4 1 7 7 2 4 8 8
1 4 5 0 9 1 1 0 0 1 5 4 5 3 3 8
0 7 8 2 0 2 0 2 0 0 6 0 1 2 7 9
2 1 3 1 4 0 5 7 4 3 6 8 4 4 3 5
8 6 0 8 1 1 4 7 2 6 2 3 9 3 1 9
1 2 9 5 9 7 8 3 8 4 3 9 4 6 1 1
8 8 1 0 0 3 5 2 4 5 6 3 8 9 1 3
3 0 1 5 2 4 9 6 9 2 1 1 2 6 3 7
5 6 5 7 7 9 4 3 0 6 5 5 1 0 1
1 5 1 8 5 1 7 3 2 3 5 1 4 8 9 1
5 7 7 5 6 3 2 4 1 3 3 8 9 6 7 2
3 5 0 0 6 8 4 6 1 9 7 2 3 3 7 2
6 0 6 4 0 7 3 2 9 3 4 4 4 3 4 3
1 0 2 3 9 5 5 1 3 4 4 5 3 8 4 0
3 0 2 3 9 9 9 1 7 2 1 0 0 3 3 7
7 7 2 6 2 6 3 7 6 3 1 0 7 8 0 7
```

- [ ] 102395
- [ ] 732643
- [ ] 251494
- [ ] 821571
- [ ] 314236
- [ ] 102127
- [ ] 121316
- [ ] 940501
- [ ] 198365
- [ ] 732351
- [ ] 648600
- [ ] 306551
- [ ] 351536
- [ ] 863475
- [ ] 544315
- [ ] 990419

Puzzle 89

```
3 9 9 8 7 0 4 9 9 4 6 5 2 2 2 5
7 0 6 4 8 0 4 0 2 7 4 1 1 8 2 1
1 9 1 9 4 5 9 3 6 6 6 4 1 8 8 0
2 9 1 8 9 9 2 2 4 6 7 3 1 9 8 8
6 6 8 0 7 9 6 7 0 6 4 8 2 0 3 1
3 6 5 2 7 0 3 6 1 6 0 9 5 1 1 0
2 3 6 8 0 7 9 6 6 9 7 9 6 7 5 4
0 1 2 0 8 8 2 1 4 3 0 1 3 1 9 7
3 7 1 1 1 2 7 2 9 8 7 8 9 8 6 4
9 6 8 9 8 7 2 1 5 5 9 5 2 0 4 9
3 7 7 1 8 3 7 4 6 7 3 7 6 0 8 9
9 9 8 9 1 3 7 2 0 3 2 6 4 4 5 7
6 9 6 7 3 5 2 8 1 3 7 3 1 3 6 5
6 4 4 0 7 6 5 3 2 9 7 1 6 2 6 1
4 0 4 3 3 6 7 0 0 1 2 3 0 4 3 3
4 0 4 6 1 4 7 8 6 1 0 4 4 6 0 8
6 1 8 0 1 5 9 3 5 2 6 8 6 8 7 6
7 1 9 7 5 2 2 4 0 6 5 4 5 2 5 9
```

- [ ] 159648
- [ ] 163710
- [ ] 979669
- [ ] 987049
- [ ] 125639
- [ ] 422885
- [ ] 533728
- [ ] 801287
- [ ] 446878
- [ ] 465736
- [ ] 396644
- [ ] 590616
- [ ] 945936
- [ ] 607697
- [ ] 437304
- [ ] 718004

Puzzle 90

```
9 7 1 7 6 7 0 7 6 3 9 6 1 0 7 8
5 0 2 8 1 5 8 0 9 6 6 1 9 3 1 3
3 3 7 7 4 1 1 0 9 8 6 5 2 7 0 7
2 8 6 4 0 4 4 0 6 4 6 3 7 7 8 5
1 2 5 8 0 1 4 1 6 1 4 5 5 9 4 7
1 8 0 1 5 2 0 8 9 9 4 6 8 2 0 0
9 3 4 4 4 6 4 0 1 3 7 1 7 2 0 1
5 3 9 5 6 9 7 9 0 2 0 6 8 6 6 3
7 1 9 9 7 8 3 1 3 3 4 2 7 3 1 7
6 3 0 2 7 8 8 8 2 5 6 3 4 6 2 1
7 9 9 7 3 2 4 1 0 2 7 6 2 2 4 3
4 8 8 1 5 9 1 3 4 4 4 2 2 5 6 7
5 2 5 7 4 5 4 6 8 5 0 1 9 7 7 6
5 5 3 5 4 3 3 6 7 4 1 0 4 4 5 8
7 2 2 6 7 1 7 6 8 6 6 8 5 3 1 1
7 4 8 4 5 0 5 4 3 8 7 0 4 0 9 2
2 3 1 3 1 9 9 3 2 7 1 9 2 0 2 6
0 7 0 9 8 0 3 1 9 9 3 4 2 5 2 5
```

- [ ] 997216
- [ ] 785570
- [ ] 440646
- [ ] 860209
- [ ] 139166
- [ ] 473419
- [ ] 804005
- [ ] 639610
- [ ] 792263
- [ ] 056721
- [ ] 483740
- [ ] 541614
- [ ] 612467
- [ ] 126988
- [ ] 392394
- [ ] 899468

Puzzle 91

```
0 5 7 8 4 5 3 1 3 0 1 9 9 9 7 7
6 1 1 2 8 7 8 6 7 3 3 0 8 7 8 6
6 3 0 1 9 5 6 1 1 4 3 1 3 7 8 1
0 8 5 4 4 0 7 7 4 7 5 3 6 3 2 6
1 3 9 1 8 0 4 3 5 6 4 4 7 9 4 9
1 2 1 9 7 4 0 4 1 1 4 4 8 4 7 1
5 8 1 4 8 8 1 9 5 9 6 2 3 7 8 2
6 4 3 4 7 3 2 9 8 9 3 6 1 2 5 9
6 2 7 0 5 3 8 2 3 2 8 0 9 6 5 4
9 2 6 7 6 2 7 3 8 1 9 1 0 1 0 0
0 2 0 8 1 4 3 8 5 4 4 1 0 9 9 2
1 4 0 0 9 9 5 4 6 3 2 5 1 3 9 4
9 6 8 1 8 8 4 9 6 8 9 5 2 1 7 6
8 5 9 8 2 0 1 6 4 9 7 7 0 8 9 6
3 7 8 1 3 3 4 2 1 5 6 8 9 2 8 1
0 3 4 5 3 9 3 2 8 2 6 7 4 7 8 0
2 6 8 8 4 1 4 1 6 0 3 5 3 7 3 6
7 0 7 6 8 9 0 2 9 4 7 6 5 5 9 0
```

- [ ] 523469
- [ ] 034761
- [ ] 635178
- [ ] 378210
- [ ] 829111
- [ ] 619823
- [ ] 173499
- [ ] 162749
- [ ] 826747
- [ ] 061414
- [ ] 901344
- [ ] 990039
- [ ] 882541
- [ ] 115669
- [ ] 678319
- [ ] 422246

Puzzle 92

```
8 5 0 4 1 3 7 9 6 0 9 5 0 2 3 2
8 3 2 7 9 1 6 5 2 2 4 1 5 6 8 3
6 2 0 6 3 8 4 8 9 4 7 8 9 2 9 4
9 9 7 0 8 8 2 7 4 3 7 7 6 3 6 3
9 1 2 7 7 7 5 9 3 4 9 2 7 2 9 3
2 3 8 0 2 0 0 2 7 2 5 6 1 3 9 2
5 7 0 6 1 9 2 1 8 9 2 0 4 5 9 3
1 3 5 1 6 7 7 5 0 3 5 1 3 2 7 2
7 4 2 4 5 6 4 7 4 6 8 9 8 7 8 1
0 7 3 7 8 0 7 1 9 4 2 7 0 7 2 8
8 6 0 7 2 0 2 6 9 9 1 7 4 8 7 3
5 5 7 9 5 8 6 8 1 6 9 7 9 7 6 9
6 6 5 6 1 4 0 1 8 0 4 1 5 4 9 4
1 4 4 7 9 9 3 7 9 9 4 3 6 8 1 6
7 3 5 4 7 4 7 6 4 9 8 3 0 4 8 2
4 4 9 0 0 6 3 6 9 7 8 2 6 3 0 7
3 7 4 2 7 4 1 0 8 1 6 7 4 9 2 6
7 3 3 1 8 0 2 0 5 7 6 0 8 2 9 8
```

- [ ] 681766
- [ ] 379609
- [ ] 197771
- [ ] 224156
- [ ] 830070
- [ ] 478382
- [ ] 981994
- [ ] 472603
- [ ] 855776
- [ ] 725613
- [ ] 346567
- [ ] 561972
- [ ] 874843
- [ ] 799969
- [ ] 734728
- [ ] 294395

Puzzle 93

```
4 8 9 8 0 5 0 8 3 5 2 6 4 8 8 0
9 9 3 5 9 3 6 8 1 6 6 1 6 6 9 7
3 4 2 9 1 2 3 0 6 9 5 5 5 1 9 0
0 6 1 1 1 8 5 0 0 4 8 9 5 9 7 1
0 7 3 7 0 6 3 3 3 7 9 5 0 8 4 5
9 8 1 4 1 2 4 6 6 7 8 6 9 0 1 2
1 7 6 2 7 1 8 6 4 4 3 0 5 0 7 5
6 6 5 7 3 6 6 4 5 3 0 9 7 4 9 9
7 4 1 5 3 4 7 3 0 3 8 6 7 9 1 1
7 7 7 8 2 7 3 8 0 1 0 4 9 6 0 2
9 4 7 5 5 7 7 9 0 8 0 7 8 9 4 0
4 1 2 7 6 9 8 5 6 5 1 9 0 0 3 6
6 8 8 8 3 1 0 0 2 7 7 9 4 5 4 5
1 0 2 9 8 7 4 0 9 1 0 6 3 7 0 7
7 9 5 9 1 2 0 6 7 0 2 4 9 9 1 1
9 6 7 9 2 6 7 1 4 4 5 6 5 7 4 2
4 2 8 7 1 4 5 7 1 8 9 6 0 9 6 0
2 6 4 3 2 7 2 4 9 5 3 3 8 1 9 6
```

- [ ] 546637
- [ ] 839164
- [ ] 176271
- [ ] 544176
- [ ] 651900
- [ ] 655095
- [ ] 846253
- [ ] 688831
- [ ] 618639
- [ ] 793730
- [ ] 210284
- [ ] 552780
- [ ] 596032
- [ ] 801049
- [ ] 805089
- [ ] 097087

Puzzle 94

```
2 2 1 9 2 7 9 1 7 9 8 0 7 4 7 2
3 0 8 0 7 6 3 8 5 7 3 2 9 2 7 7
5 7 7 1 5 8 8 4 5 1 6 0 2 2 1 7
8 1 4 3 0 3 2 9 1 0 6 7 9 0 6 6
7 5 7 8 0 1 1 3 4 2 2 5 5 3 6 1
0 7 2 5 9 2 9 9 3 6 5 7 2 9 4 3
5 7 9 7 0 4 0 2 9 6 2 4 2 6 5 3
1 3 7 9 0 5 5 7 6 9 1 7 7 0 5 5
8 8 2 0 0 5 4 9 7 5 9 7 4 2 7 0
4 8 0 4 5 6 1 5 4 0 9 4 5 2 3 5
0 0 6 2 0 3 1 3 7 4 1 5 5 3 8 3
3 6 6 0 5 6 2 2 9 1 9 3 2 6 2 9
6 7 1 4 3 4 8 0 8 3 6 6 2 1 0 7
5 6 6 7 4 7 4 5 0 1 7 1 3 3 9 4
6 1 4 1 4 6 0 1 3 6 8 8 4 0 4 2
2 4 6 8 5 2 4 9 1 4 6 8 7 3 8 6
6 7 3 9 8 0 9 3 5 8 5 8 3 0 2 2
3 1 6 3 5 9 9 6 1 2 2 7 0 2 3 0
```

- [ ] 020770
- [ ] 632206
- [ ] 341254
- [ ] 747089
- [ ] 013688
- [ ] 595629
- [ ] 676088
- [ ] 945944
- [ ] 554661
- [ ] 836621
- [ ] 927917
- [ ] 931507
- [ ] 505457
- [ ] 554722
- [ ] 864194
- [ ] 143032

Puzzle 95

```
2 5 2 7 8 5 7 0 3 1 2 1 2 0 0 5
3 7 4 5 5 0 8 0 3 0 8 6 6 2 2 3
3 2 1 2 0 5 2 5 9 3 0 8 9 6 1 5
5 1 0 6 3 9 4 7 0 5 4 7 2 1 3 8
5 7 8 5 8 9 0 0 4 5 9 2 2 9 8 5
5 5 7 3 8 3 9 1 8 3 6 8 0 5 2 2
9 1 2 5 5 9 1 5 3 1 4 8 5 5 3 7
5 4 4 8 6 0 7 7 9 5 5 2 7 6 3 5
4 3 5 9 0 5 1 7 2 5 4 9 9 9 4 7
3 3 6 1 4 1 0 6 3 4 9 9 9 7 8 4
1 5 8 3 9 3 7 6 4 0 3 3 5 0 7 9
5 1 0 2 2 1 1 7 5 2 1 6 8 9 3 6
9 8 0 0 6 7 5 5 2 7 4 0 7 9 4 2
1 8 8 7 4 1 4 5 6 8 4 7 9 8 4 7
1 4 3 0 2 4 8 5 7 8 7 8 3 4 8 5
4 6 2 0 1 3 6 2 6 4 4 1 3 4 2 3
2 7 6 5 9 7 9 0 4 4 2 4 0 0 0 6
9 3 7 2 1 0 2 8 7 1 0 2 6 5 9 3
```

- [ ] 744139
- [ ] 046739
- [ ] 716831
- [ ] 081599
- [ ] 823348
- [ ] 725576
- [ ] 462013
- [ ] 544860
- [ ] 141633
- [ ] 715095
- [ ] 587252
- [ ] 659790
- [ ] 907965
- [ ] 599324
- [ ] 865058
- [ ] 434720

Puzzle 96

```
8 2 1 0 6 2 6 6 2 3 3 5 3 8 4 3
1 8 4 0 0 5 8 2 5 8 9 6 7 9 0 2
4 5 7 4 5 9 7 8 5 4 8 6 0 0 9 4
1 8 6 3 6 2 9 7 7 2 6 4 5 9 4 0
9 2 6 2 6 6 6 0 5 5 9 5 9 3 5 1
7 8 8 0 6 9 5 3 6 9 7 3 7 4 0 7
7 6 0 8 4 5 0 7 6 8 5 7 4 4 3 8
3 6 3 2 8 5 4 0 7 0 0 4 6 0 6 6
5 4 1 2 3 3 3 9 1 7 9 9 9 6 1 2
7 2 9 3 8 3 3 9 5 8 6 8 9 1 9 0
6 4 2 2 6 1 1 5 4 8 4 3 0 2 1 2
3 7 4 2 0 0 1 6 0 9 0 4 2 1 5 1
0 8 2 2 7 2 8 2 7 5 4 4 6 5 2 1
8 4 2 0 3 0 8 6 0 4 9 8 5 7 3 3
3 6 3 6 7 7 3 2 5 9 5 1 4 8 1 7
9 3 7 4 6 0 5 5 9 4 1 2 3 5 0 3
7 6 9 4 7 7 7 6 3 2 3 0 3 4 5 1
1 8 4 4 3 3 7 2 9 5 5 1 3 7 0 3
```

- [ ] 690018
- [ ] 548494
- [ ] 642261
- [ ] 362501
- [ ] 609900
- [ ] 367537
- [ ] 288494
- [ ] 894735
- [ ] 562625
- [ ] 913086
- [ ] 519163
- [ ] 690579
- [ ] 223228
- [ ] 858286
- [ ] 064739
- [ ] 331188

Puzzle 97

```
4 2 9 4 6 4 7 8 1 0 6 6 5 3 4 2
7 3 6 1 2 4 6 6 6 8 7 7 3 2 7 3
6 6 6 5 1 5 5 4 6 1 3 8 4 3 3 0
2 5 4 1 0 1 8 9 0 0 6 8 9 4 0 6
1 7 0 2 9 8 4 2 1 7 2 5 3 3 4 6
5 6 7 4 7 5 1 1 7 8 2 0 1 3 2 5
3 5 3 7 0 0 7 4 7 9 9 3 1 0 9 9
7 9 6 9 9 7 2 1 1 1 9 6 7 7 4 9
3 6 4 9 2 1 0 6 3 8 3 2 1 4 4 0
9 5 4 8 5 5 5 0 7 1 6 8 7 0 7 1
2 3 7 7 9 2 4 1 6 1 4 2 2 1 7 0
6 8 3 6 5 5 1 0 4 4 8 9 8 2 4 1
2 1 1 3 1 0 0 9 6 7 3 5 5 9 9 5
6 7 8 9 1 5 9 7 9 3 8 9 2 5 6 0
3 7 7 9 5 9 3 3 4 6 9 9 9 7 7 6
9 9 2 0 2 1 4 7 3 7 4 5 8 6 7 5
1 1 5 2 7 0 6 5 8 7 5 9 9 7 8 4
4 9 8 6 7 6 8 8 0 1 7 9 7 2 1 5
```

- [ ] 351267
- [ ] 485550
- [ ] 258176
- [ ] 874649
- [ ] 664216
- [ ] 343307
- [ ] 936262
- [ ] 701698
- [ ] 314955
- [ ] 412029
- [ ] 065875
- [ ] 508141
- [ ] 827992
- [ ] 979389
- [ ] 854737
- [ ] 040700

Puzzle 98

```
1 0 2 3 3 6 2 3 5 0 2 6 4 0 4 2
5 6 9 6 1 7 4 9 6 9 1 6 8 5 2 5
5 8 3 9 0 7 2 4 2 9 0 1 0 3 4 9
2 5 6 8 2 7 3 3 1 4 4 8 7 1 6 1
8 2 2 6 9 7 8 7 6 0 7 4 3 4 3 1
0 9 3 2 0 7 0 3 0 2 8 5 0 1 0 9
6 6 2 9 7 6 0 7 1 4 3 8 0 6 5 8
6 5 6 6 3 6 8 0 7 5 1 9 2 7 1 8
7 3 8 8 0 7 6 5 8 2 6 9 6 6 6 8
6 5 5 2 0 1 0 5 4 1 1 3 1 1 7 5
0 4 7 2 7 1 5 0 7 3 2 5 9 0 8 0
7 4 8 7 7 3 0 6 2 8 4 3 2 7 5 2
1 3 1 1 4 0 8 6 5 0 4 2 9 0 0 9
2 2 6 2 4 3 0 7 7 2 0 1 4 9 8 9
0 4 7 4 7 5 9 2 3 4 2 0 8 1 6 1
0 0 6 3 8 8 0 7 3 6 2 9 5 8 7 6
2 0 3 3 9 9 6 9 3 9 5 0 9 0 0 7
9 2 4 2 6 3 5 3 6 5 8 6 7 7 7 4
```

- [ ] 876074
- [ ] 303117
- [ ] 987447
- [ ] 258619
- [ ] 167610
- [ ] 535443
- [ ] 692689
- [ ] 587615
- [ ] 315639
- [ ] 105820
- [ ] 187586
- [ ] 860508
- [ ] 592342
- [ ] 993218
- [ ] 924056
- [ ] 404620

Puzzle 99

```
5 1 0 0 2 4 5 7 2 0 0 5 1 9 4 9
5 6 0 1 4 6 8 5 6 4 1 5 6 7 1 2
6 5 1 7 2 5 3 6 7 6 5 1 2 7 0 7
5 3 6 3 6 9 9 3 6 3 3 3 0 5 8 8
0 6 8 4 7 6 3 1 6 7 3 8 0 9 7 3
1 9 7 1 8 4 5 5 6 6 5 5 0 2 5 5
6 2 2 1 6 1 6 7 5 9 9 7 0 3 7 2
8 2 8 5 6 5 0 1 1 0 8 0 8 8 1 5
1 9 6 7 0 7 3 8 9 1 3 6 5 9 9 0
2 6 5 0 1 6 9 0 6 3 4 0 7 3 3 6
3 3 2 8 1 2 8 8 0 6 8 2 0 2 8 4
0 7 2 5 2 0 7 0 6 5 0 4 8 4 5 6
9 8 7 7 6 0 3 4 8 2 9 5 8 4 2 9
7 5 9 3 7 0 3 3 1 3 0 3 5 5 4 7
0 5 8 3 7 3 0 7 0 9 1 5 5 8 7 3
1 7 9 4 3 8 1 7 8 2 0 2 1 2 0 8
6 7 5 9 5 6 0 5 5 7 8 6 0 3 2 3
1 5 9 3 4 4 1 0 2 5 1 4 0 4 5 1
```

- [ ] 759970
- [ ] 787302
- [ ] 268976
- [ ] 564156
- [ ] 329577
- [ ] 847631
- [ ] 084389
- [ ] 651725
- [ ] 274190
- [ ] 369936
- [ ] 256007
- [ ] 253950
- [ ] 641065
- [ ] 754200
- [ ] 610562
- [ ] 794381

Puzzle 100

```
5 0 7 1 4 6 8 2 5 0 5 6 3 5 4 4
7 2 1 8 4 0 8 3 4 0 9 8 5 0 5 0
3 9 2 5 3 7 8 6 6 8 7 4 5 7 7 6
7 3 3 3 6 3 1 1 9 6 5 5 7 1 4 8
5 3 9 0 4 3 9 8 0 9 1 2 1 4 0 4
1 4 9 7 3 8 7 2 8 7 3 0 0 6 5 3
0 6 1 7 1 3 7 5 0 1 5 1 6 9 5 5
0 0 3 8 0 4 0 6 0 0 1 3 2 1 9 3
6 5 2 7 5 5 2 4 8 8 5 4 8 8 4 1
5 5 8 9 8 7 1 0 3 9 3 3 4 5 0 6
0 5 7 2 6 3 1 5 6 1 8 9 3 5 7 2
3 5 7 6 6 4 8 8 1 2 4 8 2 9 4 1
7 6 3 2 5 0 5 0 7 2 6 6 6 4 7 8
0 5 0 9 9 8 9 5 0 7 7 6 2 2 7 8
9 6 7 8 5 3 3 7 5 1 1 9 0 8 2 9
5 9 1 4 5 6 4 6 0 4 7 2 8 0 9 4
1 9 3 9 0 6 5 0 1 4 2 5 8 0 6 0
1 1 1 5 5 5 5 7 0 6 3 1 1 4 9 2
```

- [ ] 277470
- [ ] 770358
- [ ] 083617
- [ ] 510065
- [ ] 159073
- [ ] 531579
- [ ] 211859
- [ ] 439275
- [ ] 123991
- [ ] 364310
- [ ] 304314
- [ ] 914564
- [ ] 881741
- [ ] 920053
- [ ] 710628
- [ ] 505635